The Future Is Not What It Used to Be

The Future Is Not What It Used to Be

Climate Change and Energy Scarcity

Jörg Friedrichs

The MIT Press
Cambridge, Massachusetts
London, England

MIT Press books may be purchased at special quantity discounts for business or sales promotional use. For information, please email special_sales@mitpress.mit.edu or write to Special Sales Department, The MIT Press, 55 Hayward Street, Cambridge, MA 02142.

This book was set in Sabon by Toppan Best-set Premedia Limited, Hong Kong. Printed on recycled paper and bound in the United States of America.

Library of Congress Cataloging-in-Publication Data

Friedrichs, Jörg.
The future is not what it used to be : climate change and energy scarcity / Jörg Friedrichs.
    p. cm.
Includes bibliographical references and index.
ISBN 978-0-262-01924-8 (hardcover : alk. paper)
1. Climatic changes. 2. Global environmental change. 3. Power resources—Environmental aspects. 4. Energy policy. 5. Environmental policy. I. Title.
QC903.F74   2013
333.79—dc23
2012045994

10 9 8 7 6 5 4 3 2 1

Contents

Preface

Our outlook on the future changes as time goes by, just as the horizon changes when we walk. Fifty years ago, people all over the world saw themselves moving toward different but equally prosperous visions of the future: capitalist progress and growth in the West, socialist progress and welfare in the East, and emancipatory progress and development in what we now term the Global South. In all of these twentieth-century visions, progress implied technological breakthroughs and cultural modernization, albeit partly overshadowed by the fear of overpopulation and the nightmare of a nuclear holocaust.

Since then, the panorama has changed. Not only progress but also growth, welfare, and development have been part success and part failure, delivering some of the goods but not others, and creating their own problems on the way. But none of them has come to be replaced by an alternative beacon of hope, and so despite their failings they remain indispensable props in any positive vision of the future. The nuclear holocaust has fortunately not come to pass, and overpopulation has turned out somewhat less dramatic than anticipated. But there is a growing list of other concerns, ranging from biodiversity loss and water scarcity to global pandemics and financial meltdown. The future is not what it used to be.

Today, humanity faces existential challenges that do not allow us to project past economic growth and technological progress into the future. This has implications for every single one of us. It also has serious implications for the entire fabric of industrial society, understood as the high-energy metabolic system that enabled the levels of progress, growth, welfare, and development achieved during the last couple of centuries or so.

Like the human organism, industrial society can be understood as a metabolic system because it converts inputs, such as energy and material resources, into outputs, such as goods and services, while also generating unintended side effects, such as waste heat and emissions. As a metabolic system, industrial society requires a variety of sources from which to obtain its inputs, and sinks for its unwanted side effects.

Climate change and energy scarcity are widely recognized as the most immediate choke points, with fossil fuels and greenhouse gas emissions at the heart of the matter. Climate change constrains the ability of the atmosphere and oceans to absorb our emissions "at sink" without major damage to ecosystem stability and human welfare. We therefore face a tragic choice "at source." Unless we manage to reduce the carbon intensity of the global economy, we will be forced to reduce our fossil fuel consumption, and thus stall industrial civilization, or to fry the planet, with equally dire consequences for our way of life. In the meantime, energy scarcity is also likely to result from another constraint at source: oil as the lifeblood of contemporary industrial society is becoming increasingly difficult to obtain, with serious implications for the future affordability and availability of liquid fuels.

This book takes a hard look at the prospect of our civilization coming to terms with these challenges. It precisely maps the contours of the current impasse as well as the social and political implications, and explains our manifest inability to adequately grasp and confront our predicament. Despite a healthy dose of realism, it also spells out what it would take to effectively tackle climate change and energy scarcity and thus confront the current sustainability crisis.

As a starting point, we must acknowledge that nothing in history can last forever. This applies to industrial civilization even more than to any previous scheme of social order because industrial "business as usual" is notoriously premised on growing levels of material consumption. Industrial society is transitory for the simple reason that infinite growth on a finite planet is impossible.

Climate change and energy scarcity are the most obvious cases in point. Not only is there an overwhelming scientific consensus on climate change, but there is also evidence to suggest that a steady decline of liquid fuel supply after a global peak in oil production (commonly referred to as *peak oil*) may result in serious shortages. To make a bad situation worse, climate change and energy scarcity are profoundly intertwined. As already indicated, energy scarcity may result not only from peak oil but also from a desperate need to mitigate climate change. At some point it may become necessary to restrict the consumption of abundant high-carbon fuels such as coal and unconventional oil in order to slam the brake on global warming.

Contemplating climate change and energy scarcity as key challenges to industrial civilization is tough in and of itself. Envisaging their social and political implications is even tougher. Climate change of a magnitude similar to what is currently underway has never occurred in modern history, so to understand its effects we need to hark back to historical episodes when societies were actually confronted with compa-rable stresses. The closest historical analogs to present and future climate change happened in the prehistoric Near East, especially in Mesopota-mia, and in the medieval Far North, notably Greenland and Iceland. Although these episodes are remote from modern-day experience, they show that climate change can bring societies to the brink and that their ability to recover depends on problem-solving capacity and/or social adaptability.

The situation is similar for the effects of energy scarcity. Because Western societies have never experienced disruptive energy scarcity—even the oil crises of the 1970s were only minor nuisances when compared to what may be in the offing after peak oil—we must look further afield for suitable precedents. The closest historical proxies are the massive energy scarcities that occurred in three non-Western societies: Japan, 1918–1945; North Korea, in the 1990s; and Cuba, also in the 1990s. Despite their unique features, these three cases enable us to second-guess how different parts of the world would react to a terminal decline of liquid fuel supply after peak oil. Military juggernauts are likely to seek forcible solutions to resource scarcity, while authoritarian regimes will tend to rely on the preservation of elite privileges at the cost of their own people. Only where social cohesion and communal solidarity remain strong is it possible to make sure that nobody is left behind.

Climate change and fuel depletion pose a challenge to conventional scientific modes of knowledge production because, when our entire way of life is at stake, the struggle over knowledge is bound to be political. Nevertheless, the politicization of science itself is heavily contested.

Established energy experts have for a long time retained the veneer of disinterested technocratic competence, with the peak oil community constituting a vocal counterculture. Mainstream climate scientists, by contrast, have openly embraced a transformational political role and sounded the alarm regarding global warming, with climate skeptics deploring the politicization of science as illegitimate. In energy science the alarmists remain the "cranks," whereas in climate science they have occupied the mainstream position. Because neither strategy can force society to tackle the problem itself, scientists are in a double bind: they are damned if they do and damned if they don't accept the politicization of science.

Scientific squabbles aside, the fundamental question is this: why is so little being done about climate change and energy scarcity? Despite the efforts of well-intentioned individuals and institutions, tackling such inescapable problems is hindered by three impediments that together make up a moral economy of inaction. First, people tend to greatly undervalue future events and distant strangers. The more remote somebody or something is from us, the less we care. Second, the pursuit of particular interests often thwarts collectively desirable outcomes. We may all agree that something needs to be done, and yet not a single one of us may be ready to do it. Third, there is the thorny issue of denial: people frequently treat real problems as if they were non-issues. Denial has a rational core because it minimizes pain, but it often leads to tragic outcomes.

Despite a healthy dose of pessimism—or better, realism—it is not so hard to understand what it would take to confront the current sustainability crisis. First, we would have to accept that the crisis is about our security in the most radical sense. Runaway climate change, in combination with energy scarcity, has the potential of jeopardizing not only industrial society as we know it but, along with it, our core civilizational values. But if we don't find a way to affirm our core values and preserve the essence of our way of life, everything else we care about may be lost. Second, we would have to take a variety of staunch emergency measures. Industrial society itself would have to be radically transformed to safeguard its very essence. Unfortunately, history suggests that only rarely, if ever, have such rationalist schemes been successfully implemented when the survival of a civilization was threatened by problems of its own making.

Facing the future is not for wimps. Denial and wishful thinking are for normal times, but when the going gets tough we have to let them go. This does not mean we need to despair. What we must do is take a hard look at what can be realistically expected, and brace ourselves. Whether you are a decision maker engaged in horizon scanning, a private individual contemplating investment decisions, or a citizen trying to understand what the future may hold, there is no way around climate change and energy scarcity as emerging facts of life.

Acknowledgments

This book could not have been written without the suggestions and comments of numerous friends and colleagues. It has emerged over an extended period, and I beg for pardon should I have omitted any of them from my list: Jocelyn Alexander, Mary Bagg, Sing Chew, Hugh Dyer, Matthew Eagleton-Pierce, Rosalba Fratini, Xiaolan Fu, Andreas Goldthau, Chris Hannington, Barbara Harriss-White, Eva Herschinger, Dan Hicks, David Von Hippel, Robert Hirsch, Damon Honnery, Oliver Inderwildi, Jessica Jewell, Peter Katzenstein, Sir David King, Martin Kraus, Kristian Krieger, Maite Lopez Suero, John Mathews, Rana Mitter, Paul Murray, Ruth Murray, Avner Offer, Richard O'Rourke, Gianfranco Poggi, Jochen Prantl, Stephen Quilley, Jerry Ravetz, Marcy Ross, Jörg Schindler, Hannes Stephan, Mary Stokes White, David Talbot, Marisa Wilson, Adrian Wood, Chris Vernon, and Martin Weitzman.

Special thanks are due to Clay Morgan and his colleagues at the MIT Press for their kind support, as well as the four peer reviewers for their helpful comments. Thanks are also due to the University of Oxford for granting me a sabbatical year to write this book. Elsevier, as well as Taylor and Francis, have kindly granted permission to use material from three previous articles.[1]

I would like to thank my parents for always being there, and my father-in-law for showing an interest in my work. But my deepest gratitude goes to my dear wife Kerstin and our little son Lukas who, despite many extra hours spent in the office, understood that working on this book was a passion and not simply a job.

1

The Transitory Nature of Industrial Society

There is a wry little tale about a turkey. The turkey observes that the farmer brings food every morning without fail, and so concludes that it is well provided for and has nothing to fear. As the weeks go by, it becomes fatter and maintains its comfortable view of the world until its complacency is shattered on the eve of Thanksgiving when the farmer arrives to wring its neck.[1]

The turkey in our story falls prey to what was identified by David Hume as the problem of induction: from a strictly logical viewpoint, it is inadmissible to predict future events, or infer general laws, on the basis of observed empirical regularities (Hume 2000 [1748], §4.1.20–27; §4.2.28–32).

The problem of induction appears inescapable for two categories of people: abstract philosophers in search of logically incontrovertible universal laws, and one-dimensional rationalists limited by a linear worldview like that of the turkey. For all practical purposes, however, philosophers and other sentient beings can solve the problem of induction by cultivating a critical, inquisitive mind and by taking a more sophisticated, systemic perspective.

For the sake of the argument, let us fancy an animal endowed with genuine intelligence and thus replace the "inductivist turkey" with a "reflectivist turkey." Such a smart creature could, in principle, understand the logic of the farm rather than just worrying about mealtimes. It would thereby overcome the problem of induction and understand that it is being fed only for a limited period. A systemic perspective would thus allow the turkey to grasp its existential predicament, make predictions about its fate, and perhaps even warn its inductivist fellows.

Figure 1.1
The inductivist turkey. *Source:* Courtesy of Horst Friedrichs.

This suggests a more existentialist reading of the story, to complement the epistemic interpretation. It is important to note that understanding the logic of the farm would not necessarily imply that the turkeys on that farm would be able to rescue themselves. In the worst case, they might simply lead disconsolate lives in the fearful expectation of Thanksgiving Day.

Humankind is in a similar predicament. On a finite planet, the fact that resources have been abundant in the past does not mean they will always continue to be so. As in the case of the reflectivist turkey, a nonlinear understanding of the world system mandates us to jettison the comfortable inductivist habit of extrapolating the future from the past, and enables us to make predictions about our shared trajectory. Once again, however, there is no guarantee that we are able to escape our personal and/or collective fate.

Because this is so, many people understandably prefer denial and self-deception to an unvarnished recognition of the predicament we are in. To keep deluding ourselves, all we must do is stubbornly refuse to substitute a reflectivist attitude for our inductivist mental habits. But, unfortunately, collective self-delusion does not alter the fact that, in the long run, the current industrial way of life is incompatible with planetary limits. As we will see, climate change and energy scarcity are the most obvious cases in point.

The Human Predicament

As every historian knows, any specific form of human society is transitory. Industrial society is unlikely to be an exception. It will pass away, just as all people are bound to die. Short of a miraculous breakthrough such as fusion technology, or the apparition of some other deus ex machina, industrial society cannot outlast the availability of such finite energy resources as fossil fuels. The specter of catastrophic climate change indicates that industrial society may become unviable even before the exhaustion of its resource base.

The reason is simple: in a world where material resources are finite and environmental sinks have limited capacity to absorb emissions without serious damage to natural ecosystems and human societies, the extractive and polluting intensity of industrial society is not sustainable. Once vital resources have been depleted, world population cannot continue, as it does today, exceeding the planet's carrying capacity, that is the number of people who are able to live on it sustainably at a given level of technological and economic development.[2] In the meantime, industrial society is already testing the absorptive capacity of the atmosphere via anthropogenic climate change.

Even a no-growth economy (Victor 2008; Jackson 2009) at current levels of material consumption and pollutant emissions would not be sustainable in the long run. Alas, industrial society as we know it is premised on economic growth and has a hard time dealing with recessions. Without the expectation of growth, financial markets break down and sustained investment becomes impossible. Few investors will borrow their money without the expectation of growth, since only growth enables debtors to pay back their loans with an interest.

For the sake of the argument, assume that world economic output continues to grow by 3 percent per annum. This implies that global GDP will double within twenty-three years, and quadruple within forty-six years.[3] It also implies that, a century from now *and other things being equal*, resources consumed and pollutants emitted will have increased by a factor of more than sixteen.[4] It is easy to see that such enormous growth would not be sustainable.

The obvious objection is that resources consumed and pollutants emitted can be reduced by efficiency gains and other forms of technological progress. So let us assume, again for the sake of the argument, that resource intensity and thus pollution can be reduced by a fairly ambitious 50 percent. Even so, under the above scenario, a century from now the world economy would consume eight times as many resources and emit eight times as many pollutants as today.

To continue the thought experiment, let us demand that the world economy should grow for a century by 3 percent per annum, but without any increase of resource consumption and pollutant emissions. By how much would it be necessary to abate the resource and emission intensity of the world economy (resources consumed and pollutants emitted per unit of GDP)? The answer: by a staggering 94.8 percent. To reconcile a century of 3 percent growth with the more ambitious goal of *reducing* resource consumption and pollutant emissions, the abatement of resource and emissions intensity would have to be even more drastic.[5]

Such technological miracles can certainly be imagined in theory, but how likely are they to happen in practice? Efficiency gains have been very impressive in the past, but can we simply extrapolate them into the future? Or, to revert to the specific case: How likely is the resource and emissions intensity of the world economy to sink to less than 5 percent of what it is today?

Future population levels are an important consideration. Here, the good news is that, as a result of the so-called demographic transition, world population is moving away from the historical pattern of exponential growth. Nevertheless, world population is still projected to grow by another two or three billion, from roughly seven billion in 2012 to about nine billion in 2050 and perhaps ten billion in 2100 (UN 2011; Lutz and KC 2010). Now imagine a world of nine or ten billion people where the global economy was eight or sixteen times larger than it is now. The planet would scarcely be able to yield sufficient food and essential resources, and to deal with the carbon and other emissions generated. And yet industrial society depends on the wasteful use of finite resources, as well as the capacity of the environment to absorb anthropogenic pollutants and waste products without major backlash against natural ecosystems and human societies.

One can debate the appropriateness of such back-of-the-envelope calculations, and more sophisticated arguments supported by data will be made in chapter 2. But the fundamental point is clear: high rates of economic growth may be necessary for industrial society; they may also be desirable for a variety of other reasons (Friedman 2005); and yet their effects are devastating. The problem with industrial society in a finite world is that it depletes essential resources and congests environmental sinks in an unsustainable way.

The Ghost of Thomas Malthus

None of this is radically new. The argument that exponential growth is unsustainable was already made by Thomas Malthus (1798), who suggested that growth in food production is linear and cannot keep pace with exponential growth in population levels. As a consequence, Malthus expected periodic cycles of overpopulation leading to various forms of social, political, and economic distress. This is axiomatically true if one assumes, with Malthus, that the growth of food production is indeed linear while population growth is inherently exponential.

With hindsight, Malthus got it wrong but pointed to a fundamental problem. He got it wrong because his core assumptions, which may have seemed accurate at the time, were soon to be undermined by historical developments. He pointed to a fundamental problem because, even though the specific challenge of overpopulation was to be met by various forms of progress, there are other limits to growth that can be analyzed in similar terms (Friedrichs forthcoming).

Malthus's first assumption, namely that exponential growth in food production is impossible, was proven wrong by historical developments that he could not have anticipated. From the nineteenth century, agricultural productivity was drastically intensified by industrial inputs such as chemical fertilizer and motorized machinery. Thanks to an abundant supply of such inputs, food production was largely able to keep pace with population growth. Modernists cherish the latest installment of agricultural progress as the "green revolution." From a long-term ecological viewpoint, however, industrial agriculture is problematic precisely because it has allowed population levels to temporarily overshoot long-term carrying capacity. Both fertilizers and machinery mostly rely on fossil fuels, which in turn depend on a limited resource base that is not replenished over historical timescales. Tragically, this may lead to serious population decline, or even collapse, further down the line.

Malthus's second core assumption of exponential population growth was largely correct at the time, but is much less so today. As already indicated, population growth has started to level off in many parts of the world. This is a significant reason for relief, although it should certainly not be forgotten that the world population is still projected to grow by roughly 25–35 percent by 2050 and perhaps by another 10 percent by 2100 (UN 2011; Lutz and KC 2010).

There is a third reason why Malthus's axiomatic views about cycles of overpopulation have not been vindicated by history. In line with circumstances in the early modern period, Malthus saw carrying capacity as constrained by food production at the local level. Over the last two centuries, however, mobility and trade have shifted the territorial frame of reference first from the local to the national level, then to the international, and finally to the global level. To begin with, Europeans were able to move to "underpopulated" landmasses such as the Americas and Siberia, and to import raw materials and foodstuffs from the colonies. More recently, the globalization of trade and aid have had similar effects, although in the reverse direction, buttressing population levels in developing countries. From a normative viewpoint, colonialism is often criticized as exploitative while globalization is celebrated. From a sustainability viewpoint, however, they have both contributed to a temporary extension of the world's carrying capacity, seen at the aggregate planetary level (Catton 1980).

From all this we can see how Malthus, though historically proven wrong, has pointed to the fundamental problem of limits to growth. Today these limits are of a different kind than anticipated by Malthus, and yet they pose an equally fundamental challenge to the sustainability of systemic growth patterns.

Limits to Growth

In 1972, a group of researchers around Dennis Meadows at the Massachusetts Institute of Technology adapted a neo-Malthusian framework to the planetary level and applied the method of computer-driven system dynamics to examine the earth system as a whole. In their book *The Limits to Growth* and its sequels, they have compellingly demonstrated that exponential growth in a finite world is impossible in the long run (Meadows et al. 1972; Meadows, Meadows, and Randers 1992; Meadows, Randers, and Meadows 2004).[6]

Meadows and colleagues have found that, for a while, the growth of various parameters such as world population, resource consumption, and environmental pollution may appear to defy physical limits, but only until the systemic feedbacks kick in. In the long run, as resource depletion and/or pollution exceed physical limits, an abrupt decline or indeed collapse of industrial society is the only way for the world system to return to equilibrium. The delay between temporary overshoot and ultimate collapse is due to the fact that there is a time lag between anthropogenic causes such as resource depletion and greenhouse gas emissions, and systemic effects such as energy scarcity and climate change.

The diagnosis is a systemic pattern of exponential growth, overshoot, and collapse. Contrary to what their detractors sometimes surmise, Meadows and colleagues did not envision imminent doom. On the contrary, their baseline scenario, called "standard run," displays a continued pattern of exponential growth and overshoot until about 2010, followed by the onset of systemic collapse between 2020 and 2050 (Meadows et al. 1972, 124; cf. Meadows, Randers, and Meadows 2004, 169).[7]

The end result of standard run is a contraction of the world population to the level of 1960 by 2100.[8] Shockingly, this implies a dramatic decline by more than two billion people from current levels. The decline would not happen by starvation alone, however, as it would occur over several generations and other demographic factors would also play a role: lower birth rates, pandemics, declining life expectancy driven by failing healthcare systems, and so on.

It is tempting to dismiss *The Limits to Growth* as a speculative doomsday model. But the results of the latest *Ecological Footprint Atlas*, which relies on meticulous data collection, are hardly more encouraging. According to the 2010 edition, the planet would have to be 1.5 times its actual size to sustain current levels of material consumption (Ewing et al. 2010, 18). At first glance this does not sound particularly alarming as a moderate abatement of lifestyle in the affluent world, including a change in dietary habits, would certainly make it possible to reduce levels of material consumption considerably.

The crunch, however, is the long-term constraint on food production imposed by the biological productivity of the earth. The metric of the *Ecological Footprint Atlas* is based on the assumption that non-renewable resources such as fossil fuels and other mineral inputs will always be available at present levels to prop up "bioproductivity."[9] Take away the unsustainable use of fossil fuel and mineral inputs in industrial agriculture, and feeding the current world population of seven billion people becomes impossible. With declining access to industrial inputs and increasing population pressure, plus the adverse effects of runaway climate change, the world food situation may easily spiral out of control.

As already mentioned, it is perfectly possible for industrial societies to overshoot and exceed long-term carrying capacity for a limited period of time. In the long run, however, no society, and much less the human race as a whole, can live beyond their means. No matter how recklessly we tap into the resources of the earth crust to sustain our unsustainable lifestyles, at the end of the day the increment on global carrying capacity can be only temporary.

Common Objections

There are a number of arguments skeptical readers may resort to when confronted with evidence for limits to growth. For example, they may emphasize that world population growth is slowly grinding to a halt. Due to socioeconomic and socio-cultural change, population growth has virtually stalled in many Western countries and is negative in Japan. In China it has been reduced by political intervention since 1978, although the absolute number of Chinese citizens is still rising due to demographic momentum. Population is also growing less rapidly in a number of developing countries, including Iran. As a result, world population is set to level off between 2050 and 2100 (UN 2011).

We have already conceded this and, as mentioned, it is good news from a social-ecological viewpoint. However, it is also worth recalling that world population is on track to further increase by 25–35 percent by 2050 and by another 10 percent or so by 2100. Also, the problem in today's world is overconsumption rather than overpopulation. Hence, the objection concerning the stalling of world population has some merit but unfortunately is not sufficient to invalidate the fundamental argument about limits to growth.

Skeptical readers may also reason that economic growth is not a universal feature of industrial society. After all, it is not so unusual for growth to be thrown back by economic crises. While this is true in principle, so far no economic crisis has ever altered the basic growth pattern. For example, after the credit crunch of 2008–2009 the world economy returned to real economic growth as early as 2010 (IMF 2011, 2). Moreover, the very painfulness of economic crises shows how desperately industrial society needs growth. It is true that growth is currently stalled in some affluent countries. Even in the most "postindustrial" of societies, however, a resumption of growth is anxiously awaited to reduce unemployment and prevent further meltdown of financial markets. A return to growth is invariably welcomed with relief, as happened in the United States in 2010. Meanwhile, most economic growth is now taking place in rapidly developing countries. China, for example, "requires" around 8 percent of economic growth per year, simply to maintain social stability; otherwise the aspirations of its increasingly urban population cannot be met.

Another common objection is that resource depletion and environmental pollution can be mitigated by political regulation. For example, there are fuel standards for vehicles and emission standards for industrial plants. Thus, political regulation can help with resource conservation and environmental protection. So far, however, regulation has not reversed industrial growth, nor was it ever intended to do so. On the contrary, the purpose of regulation is merely to contain excesses of wasteful resource use and to mitigate the most harmful damages from environmental pollution. Even the timeliest measures to regulate resource depletion and environmental pollution do not abrogate the fundamental unsustainability of industrial society. Moreover, there are tradeoffs between the two regulatory objectives of resource conservation and environmental protection. For example, carbon capture and storage (CCS) and other clean coal technologies reduce efficiency and thus lead to faster resource depletion.[10]

This leads us to yet another familiar objection. Will human ingenuity not lead to technical solutions for resource depletion and environmental pollution, as has sometimes happened in the past? Potential solutions include improved resource efficiency, reliance on more abundant alternate resources, revolutionary technologies such as nuclear fusion, or any combination thereof. The most sanguine technological optimists state that, while all other resources may be finite, human ingenuity is an unlimited resource that can grow exponentially forever, offsetting the depletion of finite resources and solving other problems such as environmental pollution and climate change, and maybe even transcending human biological limitations (Kurzweil 2006; Chang and Baek 2010).

It is true that, over the last couple of centuries, technical innovation has expanded the constraints within which industrial societies operate. As long as technical innovation is ahead of resource depletion and environmental degradation, industrialism offers progressive solutions. Short of a miraculous breakthrough such as nuclear fusion, however, any technological revolution is again based on finite resources. This is a serious problem because some resources such as fuel, fresh water, and fertile soils are essential in the sense that they cannot be substituted by other resources while being vital for human subsistence and/or industrial production (Ehrlich et al. 1999).

When the first essential resource has been depleted and/or when the first indispensable environmental sink has been strained beyond capacity, collapse will only be more abysmal because, short of the "progressive solutions" offered by industrialism, the carrying capacity of a materially depleted and environmentally degraded planet will be lower. This will then make a bad situation much worse (Dilworth 2010). In the unlikely event that some technological revolution gives a new lease of life to industrial society, the cycle will only turn one loop further.

There are two more reasons why one should not be overly optimistic about the ability of technical fixes to solve the human predicament. First, so far the beneficial effects of technical innovation on pollutant emission have been more than offset by growth of economic output (Dosi and Grazzi 2009). Second, it is unrealistic to assume that human ingenuity is an unlimited resource. A more realistic view is that, at the end of the day, even ingenuity is subject to the law of diminishing returns, with the lowest hanging fruit picked first. If this is so, then technical innovation is bound to become more and more challenging because of diminishing returns to investment in research and development (Strumsky, Lobo, and Tainter 2010; cf. Tainter 1988; Homer-Dixon 2006).

Some people set their hopes in a transition toward an increasingly immaterial service economy. Unfortunately, however, there is little to indicate that production can be sufficiently decoupled from resource use. Since there are thermodynamic limits to possible gains in efficiency, the economy of the entire world would have to become almost ethereal for economic growth to be sustainable in the long run (Jackson 2009, 67–86).[11] And even this would not eliminate practical limits to unlimited growth, such as the finite capacity of humans to cope with rapid innovation (Newman and Dale 2008). Besides, even the most "post-industrial" service economies rely on the import of industrial goods from rapidly industrializing countries such as China.

Bottom Line

In a world where material resources are finite and environmental sinks have limited absorptive capacity, the extractive and polluting intensity of industrial society is not sustainable. It is impossible for world population to continue endlessly, as it does today, exceeding the carrying capacity of the planet.

While this is axiomatic to neo-Malthusians, cornucopians are never going to agree. Regardless of the evidence, both sides of the debate speak incommensurable discourses (Dryzek 2005, 26–71). Cornucopians believe, per definition, that technological progress guarantees perpetual abundance. Neo-Malthusians are convinced that there are insurmountable planetary limits, not only to food production and population growth as Thomas Malthus believed, but also to the sustainability of industrialism and economic growth. To them, these limits to growth constitute an inescapable human predicament.

Cornucopians object that human ingenuity has always propelled us to the next level. They do not see any reason why this should be different in the future, and they often rely on aspirational statements. As Ronald Reagan (1985) put it in his second inaugural address, "there are no limits to growth and human progress when men and women are free to follow their dreams." If we pierce through the fog of such wishful thinking, however, and if we look at the real issues humanity is confronted with, then the neo-Malthusians have the better arguments. Infinite growth on a finite planet is impossible. Industrialism as we know it cannot last forever, and at some point industrial growth is bound to become unviable.

Industrial civilization relies on a variety of sources and sinks, some of which are more strained than others. In principle, there is no way to know for certain when overshoot will be followed by collapse, and which of many possible bottlenecks will put industrialism over the edge (Rockström et al. 2009). For all practical purposes, however, and despite unavoidable disagreement on the timing and modalities of the sustainability crisis, there is little doubt about the most likely choke points. Climate change is the most imminent strain on a sink, while energy scarcity is the most ominous resource constraint (Eastin, Grundmann, and Prakash 2011). Therefore, my central focus is on the twin challenges posed to industrial civilization by climate change and energy scarcity.[12]

2

Climate Change and Energy Scarcity

"Think globally, act locally" has always been a strange slogan. Undoubt-edly all action must happen in a place, but how can global problems be addressed unless the framework for action is also global? Local action can save the California condor and the Alabama beach mouse, but it cannot solve the planetary problem of biodiversity loss. Nor can it solve resource depletion and climate change.

For a long time, the majority of concerned people in industrial countries chose to act locally. Rather than trying to tackle global problems, they made sure that industrial societies would divert part of the wealth generated by economic growth to repair the worst effects of pollution. As a result of their noble efforts, air and water have become cleaner and many locally endangered species have survived. This, in turn, has further improved the quality of life in wealthy societies.

But alas, it has not addressed the fundamental problem: industrial growth depletes resources and generates more emissions than can be absorbed by environmental sinks without significant damage to natural ecosystems and human societies. The chicken (or turkey, in line with chapter 1) is now coming home to roost in the form of global warming. While advanced industrial societies have been successful in fixing a variety of pollution problems at the local, national, and international level, there is no effective solution in sight to fix climate change.

Recent evidence about climate change suggests that, in principle, industrial society may become unviable even before the exhaustion of its resource base. To make a bad situation worse, there are two reasons why the crisis is likely to be accompanied by physical scarcities, most notably of fuel resources.

First, industrial society can be regarded as one great source, and the planetary atmosphere as one great sink of greenhouse gases. If the sink is congested with too many emissions, there are serious consequences for natural ecosystems and human societies. Hence, climate change poses a constraint on how much fossil fuel can go up in the air. If it is not possible to reduce greenhouse gas emissions within the growth paradigm, then energy consumption may have to be rationed.

Second, there is the issue of peak oil. Cheap oil has been the lifeblood of industrial society for generations, but high oil prices are now signaling scarcity. There are substitutes such as tar sands and tight oil (commonly known as shale oil), but their production is expensive and environmentally damaging. The world production of crude oil is on a plateau and may soon enter a terminal decline. If it is true that most of the "easy oil" has been depleted, then this may pose another constraint at "source."

The consensus on energy scarcity and oil depletion is not as overwhelming as on climate change and global warming. But even the International Monetary Fund has come to acknowledge that the risk of oil scarcity is very serious (IMF 2011). The International Energy Agency seems to be oscillating between more negative and more positive interpretations of largely the same global situation from one year to the next (IEA 2011a, 2012a). The jury is still out on whether climate change and oil depletion are a direct threat to the viability of industrial society or whether they "only" pose the challenge of transitioning from the present reliance on fossil fuel in general, and oil in particular, to a more sustainable energy mix. Either way, there is broad consensus that current patterns of energy consumption must radically change not only to mitigate the effects of climate change but also to keep up with oil depletion and the concomitant risk of fuel scarcity.

I begin this chapter by briefly recapitulating the main climate change impacts: higher temperatures, altered precipitation patterns, more weather extremes, and rising sea levels. Climate change is thus pinned down as the most important and most immediate stressor to industrial society. This is followed by reflections on the climate-energy nexus. I argue that, for all practical purposes, climate change mitigation mandates a constraint on energy consumption. This is not to deny that the prospect of efficiency gains and other technological change offer some comfort. But given the magnitude of the challenge and the trajectory we are on, it is unrealistic to expect that technological change alone can solve our problems.

This brings me to energy scarcity, and especially arguments about peak oil. According to these arguments, there is a serious risk that oil depletion may lead to an escalating scarcity of oil, which remains the lifeblood of late-modern industrial civilization. Subsequently, my focus shifts to how oil depletion and climate change can be mitigated by a modified energy mix of fossil fuels, nuclear power, and renewable resources. Non-oil fossil fuels, namely coal and gas, are fairly abundant, but it is technically difficult and economically expensive to substitute them for oil. Especially for coal but also for non-conventional oil, there is another serious problem: if these high-carbon energy sources are substituted for conventional oil, the effects on climate change will be disastrous. Nuclear energy has its own problems and cannot make up for the shortfall. Renewable energy is clearly the most desirable policy option, but the technical and economic obstacles to significantly expanding its share in the global energy mix are daunting.

This leaves industrial society caught between the hammer of climate change, mandating restraint on fossil fuel consumption, and the anvil of peak oil, mandating an increased reliance on fossil fuels that are more abundant than conventional oil, including the most abundant and most carbon-intensive of all fossil fuels: coal. Bold visions about a low-carbon or even zero-carbon economy are precisely that: bold visions—a point underlined in the final section, "The Energy Quagmire."

Climate Change

There is no need, for our present purposes, to delve into the complexities of climate science. Instead, the discussion can be limited to the most important manifestations of climate change: higher temperatures, altered precipitation patterns, more weather extremes, and rising sea levels. These impacts are well known (IPCC 2007b, 2012; see also Houghton 2009, 135–238; Archer and Rahmstorf 2010, 125–190; Dessler 2012, 136–152), but it is worthwhile to recapitulate them as a baseline for further discussion.

The most important point to bear in mind is that human civilization has co-evolved with the relatively stable climate of the last 10,000 years.

Our cultural practices, built infrastructure, and economic capital are all designed for that climate. While in principle humans are able to survive under a variety of climate regimes, any rapid change will pose huge challenges to our adaptive capacity.

Higher Temperatures
At the end of the twenty-first century, average global temperatures are projected to be roughly 2–7°C above preindustrial levels. This can be decomposed into about 0.6°C of global warming from about 1750 to 1990, plus an additional 1.1–6.4°C in the period from 1990 to 2100.[1] Overall, it seems safe to say that the world must be prepared for global warming of at least 2°C, and probably more, above present temperatures by the end of the twenty-first century.

It is important to note that even a warming of only 2°C would lead humankind into dangerously unchartered territory. The world is currently about as warm as during any other interglacial period of the last two million years, and about five or six degrees warmer than twenty thousand years ago during the last ice age. If global warming turns out at the upper limit of the projections, the temperature increase since the last ice age will be repeated over the next century. Many animal and plant species are not adapted for such radical change and would face extinction. As a consequence, entire ecosystems would collapse.[2]

Within the general trend of global warming, there is expected to be significant geographic variation. As a general rule of thumb, warming is expected to be higher over continental land masses than near the coast or over oceans. It is expected to be particularly high in the Arctic region where it may reach about twice the global average, except for a spot in the North Atlantic Ocean where warming is likely to be offset by a slowdown of the Gulf Stream.

There is a hard-nosed view in some quarters that there are going to be winners and losers from climate change. For example, some observers are speculating that the boreal parts of the Northern Hemisphere stand to benefit from global warming in terms of agricultural output and human habitat. Thus, vast swaths of Canada and Siberia, as well as Greenland and Iceland, have been declared climate-change winners in a recent imaginative account (L. C. Smith 2011). While in this scenario the Far North does not have to worry about climate change, populations in the Global South are likely to bear the brunt. As has been usual since the onset of industrialization, the most vulnerable places are said to be in the tropical region and most notably in sub-Saharan Africa.

The problem with this narrative is that it works only if we understand climate change as limited to global warming. In reality, however, this is not the case. As we shall see in a moment, climate change is also associated with erratic precipitation patterns and more frequent, extreme, and unpredictable weather events. Furthermore, rising sea levels must be added to the equation.

Altered Precipitation Patterns

While intuitively one might expect a hotter planet to be drier, this intuition is wrong. On the contrary, overall levels of precipitation are largely going to increase because, on a hotter planet, more water will evaporate and then come down again as rainfall. Rainfall changes are extremely important not only to ecosystems but also to human societies, which undeniably need access to secure supplies of fresh water not only for drinking and sanitary purposes, but also for agriculture and thus food production (Oki and Kanae 2006).

As with rising temperature, there is a hopeful but naive view holding that a global increase in precipitation should improve secure access to water. Unfortunately, once again this is not the case. While it is true that climate change will increase the total amount of fresh water available to human societies and natural ecosystems, on the local and regional level precipitation patterns will become significantly more unequal and erratic. Rainfall is expected to increase in areas that are already humid, such as moist tropics and high latitudes like Siberia and northern Canada. In relatively dry mid-latitudes such as the Mediterranean, as well as semi-arid low latitudes such as southern Africa, by contrast, precipitation levels are expected to decrease. In humid mid-latitudes, such as the northern part of Europe, the climate is expected to become drier in summer and wetter in winter, leading to a higher risk of undesirable summer droughts and winter floods.

On balance, the number of people living in areas suffering from water stresses is likely to increase.[3] Especially in arid and semi-arid lands, reduced access to water will not only curtail drinking water for many people, but will also put considerable strains on food production. Overall, the changes in precipitation patterns are expected to lead to serious adverse consequences.

More Weather Extremes

If it were only for an increase in mean temperatures and precipitation levels, then future climate would simply be characterized by a higher frequency and intensity of hot and wet weather. This would be counterbalanced by a lower frequency and intensity of cool and dry weather. While such changes would be of most concern to people living in places that are already hot and wet, others might find them relatively easy to adjust to.

Alas, as we have already seen, climate change is associated with significant geographic variation. In addition, it also comes with a greater frequency and intensity of extreme weather events, while "normal" weather becomes less pervasive. Thus understood, climate change must be of concern to virtually everybody as few people on this planet will be exempt from more frequent and more intense droughts, floods, heat waves, and perhaps even cold spells.

This is illustrated by figure 2.1, where the shaded area symbolizes the situation before climate change, with its specific probability distribution of "normal" and "extreme" weather, while dotted areas symbolize various alterations related to climate change. The first panel stands for an increase in the mean, while the second panel stands for an increase in the variance of future climate. The third panel shows the combined effect of an increase in mean and an increase in variance: extremely hot and wet weather becomes far more frequent and extreme, whereas very cold and dry weather continues to occur, although perhaps not with the same frequency.

From a human perspective, extreme and unpredictable weather is even more worrying than a general increase in temperatures or precipitation levels. This includes, but is not limited to, droughts and heat waves. There will also be a higher frequency and intensity of storms, due to changed wind conditions. As a result, storm surges and tropical cyclones must be expected to become more frequent and hazardous as sea levels rise.

Rising Sea Levels

Rising sea levels are another adverse effect of climate change. In many parts of the world, low-lying coastal plains are particularly densely

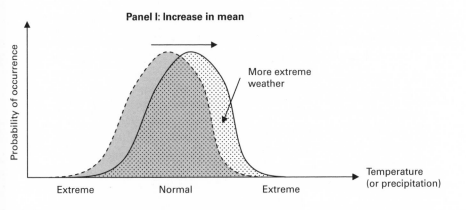

Panel I: Increase in mean

Probability of occurrence

More extreme
weather

Temperature
(or precipitation)

Extreme   Normal   Extreme

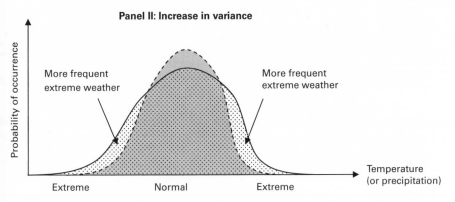

Panel II: Increase in variance

Probability of occurrence

More frequent
extreme weather

More frequent
extreme weather

Temperature
(or precipitation)

Extreme   Normal   Extreme

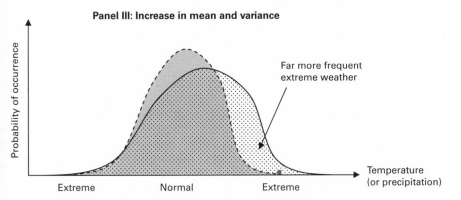

Panel III: Increase in mean and variance

Probability of occurrence

Far more frequent
extreme weather

Temperature
(or precipitation)

Extreme   Normal   Extreme

Figure 2.1
Normal vs. extreme weather. *Source:* Adaptation of Figure 2.32 in *Climate Change 2001: The Scientific Basis. Contribution of Working Group I to the Third Assessment Report of the Intergovernmental Panel on Climate Change*, Cambridge: Cambridge University Press, p.155. See also the similar figure in IPCC 2012, p.7.

populated. In total about 13 percent of the world population live in endangered coastal areas, at less than 10 meters of altitude (K. Smith 2011). The main problem does not emanate from rising sea levels in and of themselves, but from the fact that climate change will increase the impact and risk of storm surges and tropical cyclones. Even today, storm surges and tropical cyclones are notorious for jeopardizing valuable human habitat and the fertility of intensely farmed areas.

Especially in developing countries with important but largely unprotected river deltas, such as Bangladesh, Vietnam, or Egypt, a rise of sea levels by only half a meter would add enormous pressures. On low-lying small islands, climate change has the capacity of wiping out entire nations and their cultures. Although there are relatively few people living on low-lying islands such as the Maldives, at least when compared to major river deltas, the loss of human habitat would obviously hit the beleaguered islanders particularly hard.

There are two main reasons why sea levels are rising: First, water volume expands with increasing temperature (and decreasing salinity). Second, oceans receive additional water from the melting of land-based ice. While thermal expansion is fairly predictable, there are considerable uncertainties related to the large ice sheets of Greenland and Antarctica. In principle it would take thousands of years for these to melt, but the pace of sea level rise will dramatically increase should they "slide" into the ocean. Unfortunately the physics of ice sheet dynamics is poorly understood. This especially applies to the West Antarctic Ice Sheet, which sits on bedrock below sea level, making it particularly vulnerable to melting from below due to warming sea water.

The IPCC estimates are very conservative, with only 18 to 59 centimeters of sea level rise predicted before the end of the twenty-first century. Since the appearance of the 2007 IPCC report, there has been increasing talk among scientists about 1 meter of sea level rise, and maybe more (Manning 2011, 43–45). If one takes into account the possibility of ice sheets sliding into the ocean, then sea level rise in the range of 1 or 2 meters becomes conceivable (Archer and Rahmstorf 2010, 139–144).

Other Climate Impacts
There are other climate impacts apart from higher temperatures, altered precipitation patterns, more weather extremes, and rising sea levels. They

include, but are not limited to, a reduction of sea ice cover in the Arctic; melting mountain glaciers; and ocean acidification. Some of these impacts are potentially beneficial. For example, a reduction of sea ice cover in the Arctic would benefit resource extraction and sea commerce. Others are unequivocally negative, such as the threat posed to marine ecosystems by ocean acidification.

The net result of climate change impacts will be negative, and there are unlikely to be any long-term winners. While there may be relative winners in the sense that some areas (notably in the North) will suffer less from the effects of climate change than others (notably in the South), it is highly unlikely that any region or country will be better off with climate change than without.

Contrary to the rosy scenario depicted by some (e.g., L. C. Smith 2011), even the Far North will suffer from more weather extremes, rising sea levels, and the risk of radical discontinuities. Polar agriculture is extremely susceptible to soil erosion. Although the effects may be positive in some locations on some occasions—for example, in Siberia in a year without either flood or drought—they will be negative in most places most of the time. Should the Gulf Stream be interrupted, the shores of the North Atlantic may experience a sudden switch from warming to cooling.

On balance, climate change is a significant stressor to any human society. While the adaptability of ecologically marginal habitats is tested far more severely than the adaptability of ecosystems on large humid landmasses, no human society should expect to benefit from climate change. What matters is the adaptability of social and political systems, namely how vulnerable and/or resilient they are and whether they are going to be stressed beyond their problem-solving capacities.

More and more people will suffer from water stresses; many areas will be affected by reduced crop productivity and worsening food shortages; there will be higher risks to human health from various diseases; heat waves and other extreme weather events will lead to human fatalities and damage to property; numerous plant and animal species will face extinction; entire ecosystems will collapse.

Note that the climate models of the IPCC do not assume any "tipping elements" (Lenton et al. 2008), or radical systemic discontinuities. While a slowdown of the Gulf Stream is factored into some of the climate models, none of them considers a full-blown shutdown of the North Atlantic Oscillation (NAO). Nor does any of the IPCC's climate models consider a radical transformation of monsoon patterns, a collapse of the Amazon rainforest, or a cascading release of the climate gas methane due to the melting of arctic permafrost. Remarkably, this is despite the fact that scientists sometimes estimate the likelihood of such "low probability high impact" risks as high as 10 percent (Archer and Rahmstorf 2010, 146).[4]

This raises the question of how much risk humanity is willing to take before taking action. Most of us would certainly be willing to buy insurance against a 10 percent risk of our house burning down, and a nuclear plant would never be built if the risk of a serious accident were 10 percent. As our failure to prevent climate change suggests, however, such precaution does not apply to the planet.

The Climate-Energy Nexus

The most important direct policy implication of climate change for the energy system is the need for carbon abatement. Humanity simply cannot stay on its current emissions trajectory, or the consequences for climate change will be dramatic. By February 2013, the world had already reached a level of more than 396 parts per million (ppm) of CO_2 in the atmosphere.[5] Although this does not include carbon-equivalent values for other greenhouse gases such as methane and nitrous oxide, it is disturbingly close to the 450 ppm commonly associated with a long-term global warming of 2°C. At present rates, another 2 ppm is added to the balance every year. This means that, unless strong action is taken very quickly, global warming exceeding the agreed-upon tolerance limit of 2°C is virtually unavoidable.

The growth pattern in CO_2 emissions has been consistent over the last fifty years (see figure 2.2). It is important to note that the general trend has not been abrogated by any economic crisis during that period. While there have been periods of moderate carbon rebate after the oil crises of 1973 and 1979, after the collapse of the Soviet Union in 1991, and after the Asian financial crisis of 1997, in no case has the effect lasted for more than three or four years. Subsequently, carbon emissions have invariably returned to their growth trajectory (Peters et al. 2012).

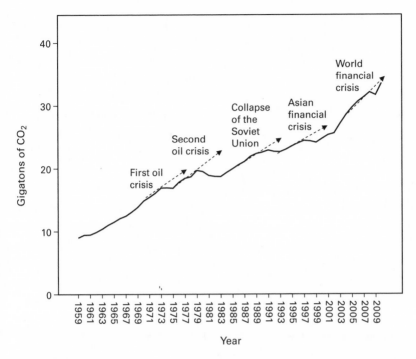

Figure 2.2
Global CO_2 emissions. *Source:* http://www.tyndall.ac.uk/global-carbon-budget
-2010.

Emission growth in the 2000s was 3.1 percent per year. This
was higher than in the 1980s and 1990s.[6] Similar to other crises, the
2008 global financial crisis had an impact, but it was weak and lasted
for only one year, namely 2009. Due to rapid economic growth in developing countries, most notably China, a carbon rebate of 1.4 percent in
2009 was followed by an unprecedented carbon rebound of 5.9 percent
in 2010. In absolute terms, the year 2010 has seen the highest total
annual growth of CO_2 emissions ever recorded (Peters et al. 2012).
Despite serious economic problems in Europe and other countries of the
Organisation for Economic Co-operation and Development (OECD),
world carbon emissions increased by another 3.2 percent in 2011 (IEA
2012a, 68).

Not all CO_2 emissions are from fossil fuels, and CO_2 is not the only
greenhouse gas. But most greenhouse gas emissions (about 60 percent)
do come from fossil fuels, and no other greenhouse gas is as important

and durable as CO_2. Thus, for our purposes it is not really necessary to complicate the picture by adding CO_2-equivalent greenhouse gases and subtracting non-gaseous forcing agents such as particulates due to their cooling effect. It might even be misleading as current emissions of most non-CO_2 agents will be long gone when most of the CO_2 currently released to the atmosphere will still be there, warming the planet.

There is an international consensus that, because global warming beyond 2°C is seen as very dangerous, CO_2 emissions into the atmosphere must be drastically reduced to prevent global warming beyond 2°C. But current patterns of energy use are manifestly incompatible with such climate change goals. According to the International Energy Agency (IEA 2012a, 52), a continuation of current emission patterns would lead to an incredibly dangerous temperature increase of 5.3°C. In contrast to this "Current Policies Scenario," the IEA's "New Policies Scenario" assumes compliance with declared policy intentions, including cautious implementation of the Copenhagen Accord. But even this more optimistic scenario would leave the world on track toward long-term global warming of 3.6°C. While 3.6°C of global warming is certainly preferable to 5.3°C, even a temperature increase of 3.6°C would be enormously dangerous.

As indicated, experts and policy makers largely agree that any global warming beyond 2°C would lead to unacceptable risks. Therefore, the IEA has developed the "450 Scenario" (IEA 2011a, 205–242; 2012a, 241–265). The 450 Scenario works backward from the goal of limiting global warming to 2°C, which corresponds to an atmospheric CO_2 concentration of 450 ppm. The stated purpose of the exercise is to identify technologically and politically plausible pathways to reach that goal without seriously disrupting industrial civilization.

The IEA suggests that the 450 Scenario is compatible with a 16 percent increase in global energy consumption by 2035. By then the world would burn 33 percent less coal and 10 percent less oil, but 21 percent more gas. The share of fossil fuel in the global energy mix would decline from 81 percent to 62 percent, while the share of renewable energy would rise from 13 percent to 27 percent. The share of nuclear power would also rise significantly, from 6 percent to 11 percent of all energy produced (IEA 2012a, 241–250, 552–553).

The bottom line of the 450 Scenario is that the 2°C goal can be met while increasing energy consumption, and without immediately abandoning fossil fuels as the main source of energy. It would leave the global economy free to consume all available fossil fuel reserves, although at a slower pace. This sounds reassuring, were it not for a number of problematic assumptions.

First and foremost, the IEA makes enormously bold assumptions about the possible reduction of carbon intensity, defined as CO_2 emissions per unit of GDP. In its 2011 *World Energy Outlook*, the agency was fully aware of this boldness: "CO_2 intensity . . . declines at a rate of 3.5% per year from 2010 to 2020, and 5.5% per year from 2020 to 2035. This rate is more than six-and-a-half times greater than the annual intensity of improvements achieved in the last ten years" (IEA 2011a, 213). Since then, the position has become less transparent. In 2012, the agency complicated its own working definition of CO_2 intensity and failed to provide a full specification of the assumed trajectory (IEA 2012a, 252).[7]

The problem is that, while drastic carbon abatement is easily factored into mathematical models, it is very hard to accomplish in the real world. This is not to deny that, as can be seen from figure 2.3, the global economy has become considerably less carbon-intensive over the last forty years. During the 2000s, however, the decline curve has flattened to a meager 0.77 percent per year.[8] This is particularly disturbing if we consider the fact that the 2000s, unlike previous decades, have seen considerable investment in carbon abatement.

While carbon abatement is apparently becoming more difficult, the IEA assumes that the trend can be reversed by political fiat. Figure 2.4 plots the improvements in carbon intensity assumed under the 450 Scenario of the IEA (2011a) against the real historical trend observed between 2000 and 2010—when carbon intensity declined by 0.77 percent per year, as mentioned. It is easy to see that the 450 Scenario would require enormous investment in green energy policy, and/or the successful introduction of major new technologies such as carbon capture and storage (CCS), to reverse the trend line and get the world on the trajectory envisaged by the IEA.

The IEA estimates the additional investment required under the 450 Scenario at $16 trillion by 2035, in 2011 US dollars. This sum, which

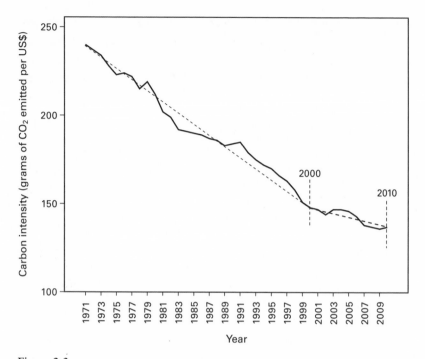

Figure 2.3
Carbon intensity of the world economy, 1971–2010. *Source:* Peters et al. (2012); dataset courtesy of Glen Peters.

would come on top of the investment already required under the less ambitious New Policy Scenario, amounts to roughly one US annual GDP (IEA 2012, 252).[9] The estimate seems extremely conservative if we consider that the abatement cost is likely to rise exponentially as carbon intensity approaches zero, and that policy makers are likely to be guided as much by political concerns as by efficiency considerations when it comes to deciding which specific carbon abatement policies to adopt.

Another challenge is the problem of sunk cost, or carbon lock-in. Unless we assume that the world's existing capital stock will be systematically retrofitted or that carbon-inefficient power plants, factories, buildings, and the like, will be decommissioned during their operating lifetime, already existing infrastructure locks in 81 percent of the "budget" of CO_2 emissions that are permissible under the 450 Scenario until 2035. If present trends continue, carbon lock-in is on track to rise to 100 percent by 2017. This means that the infrastructure installed by 2017

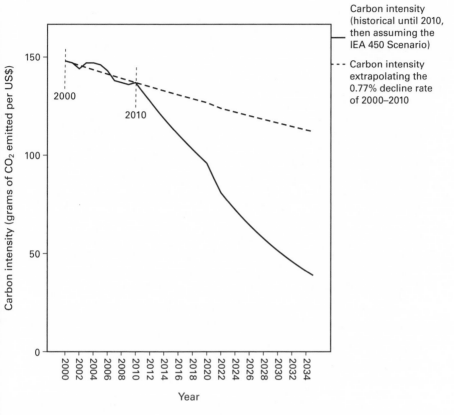

Figure 2.4
Carbon intensity of the world economy, 2000–2035

will, due to its lifetime, "commit" the world to atmospheric greenhouse gas concentrations of 450 ppm of CO_2-equivalent. If 450 ppm is to be the ceiling, any new carbon-emitting infrastructure would have to be offset by a reduction of carbon emissions from existing infrastructure (IEA 2012a, 261–265). Such early retirement of existing infrastructure would be enormously expensive. In its 2011 report, the IEA noted that it was "theoretically possible at very high cost, but probably not practicable in political terms" (IEA 2011a, 205). In its more upbeat 2012 report, the agency silently dropped this important disclaimer (IEA 2012a).

A third challenge is carbon capture and storage. In 2011, the IEA assumed that CCS would become viable on a large scale between 2020 and 2035, contributing 22 percent of carbon abatement during that period (IEA 2011a, 214). The problem is that the prospects for CCS remain very uncertain. As the IEA admits, CCS is beginning to emerge much more slowly than anticipated and may not be available on any significant scale before 2030 (IEA 2011a, 236–242; 2012b, 42). In fact, the agency has downgraded its contribution to 17 percent in 2035 (IEA 2012a, 253).

A drastic reduction of carbon intensity, an instant halt to carbon lock-in, and rapid deployment of CCS—these are just three of the heroic assumptions that would have to come true if the 2°C goal were to be compatible with familiar patterns of energy consumption and economic growth. But how realistic is all of that?

When fed with the "right" equations, a computer will produce anything you want. For example, the IEA envisages a zero emissions scenario for 2075, but admits that this would require implausible technological breakthroughs (IEA 2012b, 513–533). Maybe if the whole world were a research laboratory run by exceptionally gifted engineers following the advice of enlightened economists, a genuine energy technology revolution would be feasible. In the real world, however, such hopes are a technocratic fantasy at best and an irresponsible distraction at worst.

Outside the happy world of entirely hypothetical modeling exercises, tackling climate change seems impossible without a politically enforced curtailment of fossil fuel consumption. There is no way around the choice of intentionally burning much less carbon even where it hurts, or frying the planet.

From an ethical viewpoint, burning significantly less fossil fuel is the right thing to do. From an economic and political viewpoint, however, this is not where the problem ends but where the problem begins. A political decision to seriously curtail fossil fuel consumption would send shockwaves through the economy. There is three times as much carbon in existing fossil fuel reserves as can be emitted under the 450 scenario (IEA 2012a, 259). In a way, this is unburnable carbon. But since the economic growth engine requires continuously high inputs of energy (Warr and Ayres 2010, 2012), industrial society cannot afford to go "cold turkey" on fossil fuels. Besides, some of the most important publicly listed companies hold enormous reserves of unburnable carbon as assets on their balance sheets. A burst of this carbon bubble might lead to a serious crisis, including financial meltdown at crucially important stock exchanges such as London (Carbon Tracker Initiative 2011). So we are stuck with the problem.

But maybe geoengineering can bail us out? Some authors are seriously discussing desperate solutions such as adding carbonate to the oceans or injecting aerosols into the stratosphere (Brown and Sovacool 2011; Moriarty and Honnery 2011). Alas, there are two reasons why this is unlikely to work. First, there will never be a global consensus on specific measures. For example, India will oppose anything that might interfere with monsoon patterns while Russia may not agree that global warming is a serious problem in the first place. Without a consensus, however, unilateral interventions will either cancel each other out or cause unpredictable compound effects. This leads us to the second problem. If we barely understand the climate system well enough to grasp what we are doing to our planet, how can we possibly hope to control the effects of geoengineering experimentation?[10]

From a thermodynamic viewpoint, systems can be constrained at source and/or at sink. So far we have primarily dealt with a constraint at sink, notably in the atmosphere—which cannot absorb escalating carbon emissions without major damage to natural ecosystems and human societies. We have seen that, for all practical purposes, climate change mitigation requires a solution at source, notably a considerable curtailment of the consumption of fossil fuels. The resulting energy scarcity would have severe consequences for industrial society. In a way, industrial civilization would be committing suicide for fear of death.

Energy Scarcity

To make a bad situation worse, thermodynamic constraints may occur not only at sink but also at source. If the world runs out of fuel, climate change will run its course due to the carbon accumulated in the atmosphere. But the climate situation would ultimately stabilize, while the real threat may be energy scarcity. This is precisely what has been suggested by dissident scholars (Höök and Tang 2013) and popularized by radical observers (Heinberg 2007, 2011). These people claim that, regardless of climate change, the world is about to face disruptive energy scarcities on the supply side, due to the imminent peak of virtually any non-renewable energy resource from oil to gas, and from coal to uranium.

On closer inspection, most of these forebodings fall apart. Some of the more extreme emission scenarios may be overblown, although for good or for ill fuel reserves are sufficient to fry the planet (Ward et al. 2012; Helm 2012). But in one case there is genuine reason for concern—namely oil. This is highly unfortunate as, for the last few generations, cheap and abundant oil has been the lifeblood of industrial society. It will not be easy to find suitable substitutes for this liquid fuel in key economic sectors such as transportation and petrochemistry.

Peak Oil

The Stone Age came to an end not for a shortage of stones. The Coal Age came to an end not for a shortage of coal. But, contra former Saudi oil minister Sheikh Yamani (quoted in M. Fagan 2000), the Oil Age may come to an end for a shortage of oil. This is what "peak oilers" are suggesting. They claim that oil production is currently peaking and will soon start a terminal decline. Most of them imply, further, that no adequate alternate resource and technology will be available to replace oil as the lifeblood of industrial society.[11]

Oil is the lifeblood of industrial society in three different ways: at 32 percent, it constitutes the largest share in the world energy mix (IEA 2012a, 552); it is vital for the smooth functioning of traffic and transport, which in turn are indispensable for a smooth functioning of economic processes; and, due to its unique qualities as a high-density liquid fuel, adequate substitutes are very hard to come by.

The last point in particular warrants some further elaboration. As the International Monetary Fund has found (IMF 2011, 93–97), the price elasticity of oil is extremely low. According to the IMF's estimates, which are based on the period from 1990 to 2009, a doubling of oil prices would lead to an immediate reduction of demand by only 2 percent. Over twenty years, the effect would grow to a still relatively meager 7 percent. The price of a barrel would have to increase as much as sevenfold, to about $700 (inflation-adjusted), for oil demand to be 50 percent lower than it would otherwise be, after a period of two decades.

So much for the theory; in practice, the economy would hardly be able to absorb an oil price hike to $700. While the 2008 world economic crisis was caused by a variety of imbalances, the escalation of the oil price to almost $150 arguably was an important contributing factor

(Hamilton 2009, 2011, 2013). As long as the oil dependence of the world economy remains unbroken, there is likely to be a global crisis whenever the oil price goes through the roof, with the ceiling somewhere between $150 and $200 in 2010 US dollars (Lipson 2011).

The main reason for oil's low elasticity of demand is that most of the potential for substituting oil by other resources was exploited in the 1980s, in the aftermath of the energy crises of the 1970s. Today, oil is mostly used in transport and in the petrochemical industry, where developing adequate substitutes is very costly and requires considerable lead times. For example, electric cars offer some hope for the future, but replacing the existing vehicle fleet would require enormous investment and take decades (Hirsch, Bezdek, and Wendling 2005, 2010).

Running Short?

Because substitution has become so difficult, peak oilers are credible when they claim that no adequate alternate resource and technology will be available to replace oil as the backbone of industrial society. But are they equally justified in their main premise: namely, that oil production is about to decline and that this must lead to an escalating scarcity of liquid fuels?

After interminable debates on the mathematical form of the curves by which oil production first rises and then declines (Brandt 2007, 2010), even the International Energy Agency now acknowledges that ever since the period of 2005 to 2008 the world production of crude oil has been on an undulating plateau. The IEA claims, however, that crude oil production will not fall much below its current level until 2035, and that a combination of unconventional sources may allow for a moderate but steady increase of world oil supply to about one hundred million barrels a day (figure 2.5).[12]

Peak oilers are less sanguine. They propound four reasons why crude oil production cannot stay for long on this plateau, and why the world supply of liquid fuels will soon enter a rapid decline. First, there is a runaway decline in output from existing oil fields. Second, unconventional oil and alternate liquid fuels are already struggling to compensate for the decline. Third, energy return on energy investment (EROI) is declining. Fourth, there is not enough oil "yet to be found" and "yet to be developed."

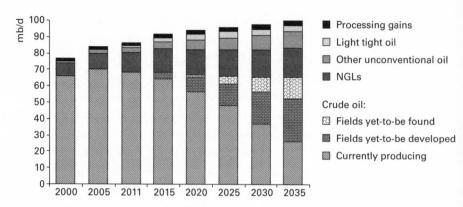

Figure 2.5
World oil supply in the New Policy Scenario. *Source: World Energy Outlook 2012* © OECD/IEA, Figure 3.15, page 103.

Let us begin with the declining output from existing oilfields. While the magnitude of oil reserves is important in the long run, what matters in the short term is the amount of oil that gushes from the ground. Since markets depend on "flows" rather than "stock" (Fantazzini, Höök, and Angelantoni 2011, 7867), it is crucial to look at decline rates. Most of the currently producing oil fields have already passed their peak, and their production is declining at a rate of 5–6 percent every year. If one includes the minority of oil fields that are still in the build-up, then this translates into an overall decline rate for world crude oil production of about 4–4.5 percent per year (Sorrell et al. 2012)[13].

The implications are daunting. According to the IEA (2012a, 102), crude oil production from fields that were producing in 2011 will drop by almost two-thirds by 2035. This means that, by 2035, capacity additions of 40 million barrels a day will be needed just to compensate for declining production at existing fields—more than the current production of all OPEC countries combined.

Most of the shortfall is currently made up from resources that are far more difficult and far more expensive to produce, such as deepwater oil from the Gulf of Mexico, or "unconventional oil" such as tar sands from Alberta or extra-heavy oil from Venezuela. The latest addition is light tight oil from the United States, also called shale oil. While deepwater oil is notoriously challenging, production from unconventional sources is equally constrained. A good example is the Athabasca tar sands from Alberta, Canada, the production of which requires enormous amounts of gas and water. Another example is the production of biofuels, which is competing with food production. Since the marginal production cost of these alternate liquid fuels is high, the industry is obviously not turning to them voluntarily but because the era of cheap and easy oil is over.

A significant problem with most liquid alternatives to conventional crude is declining energy return on energy investment. EROI is defined as the ratio between energy output from an extraction process and the energy input required to fuel that process (*output/input*). The current EROI of oil is somewhere below 18:1, down from about 25:1 around 1970 (Cleveland 2005). It is still higher for crude oil from places like Saudi Arabia, but it is considerably lower for unconventional sources such as deepwater oil and tar sands. Declining EROI implies that increasing shares of the energy produced have to be reinvested in energy production. Moreover, below an EROI of about 10:1 the effects on market prices become disproportionately more significant (Heun and de Wit 2012).[14]

Finally, it is doubtful whether the two wedges of "yet to be found" and "yet to be developed" in figure 2.5 can be as large as expected by the IEA. The agency assumes that even in 2035 most of world liquid fuel production will still be coming from conventional crude, but this depends on enormous amounts coming online. The crux is that little "easy" crude oil is being discovered. Gone are the days of the "black gold" that popular imagination has gushing out of the desert. Peak oilers point out that, since the early 1980s, more oil has been produced every year than was discovered (e.g., Sorrell et al. 2010b). They claim that there is simply not enough oil "yet to be found" and "yet to be developed" to avert a serious decline in oil production.

Despite its apparent plausibility, the claim of peak oilers that global oil reserves are shrinking is actually their weakest argument. One problem is that, apart from new discoveries, there is at least one other legitimate reason why reserve figures are constantly augmented: improvements in technology or higher oil prices can make the production of formerly unprofitable oil deposits economically profitable, thus leading to their reclassification. Thus, it is not unusual for previously inaccessible *resources* to become recoverable *reserves*.

Another problem is that it is hard to tell legitimate from illegitimate reasons for reserve growth. To be sure, OPEC countries have an incentive to massage their reserve data to boost their production quota, and shamelessly did so in the 1980s. Similarly, international oil companies have an incentive to inflate the reserves listed on their balance sheets to look stronger on the stock market. But absent a proper verification mechanism, we can never know for sure. The upshot is that data on global oil reserves are too uncertain to offer any practical guidance (Owen, Inderwildi, and King 2010; Sorrell et al. 2010b; Sorrell et al. 2012).

And yet we can be sure that oil is becoming scarce, simply because of the combination of declining production from existing oil fields and the manifest difficulty of finding economically attractive substitutes. The petroleum industry is struggling to keep the production of crude oil on its current plateau despite the fact that oil prices have been unusually high over most of the last ten years. If oil was not becoming scarcer, the market would have reacted to higher oil prices with more supply. Despite the recent surge of shale oil in the United States, this is not happening at the global level. An inflexible supply is struggling to meet an inelastic demand, so that "the era of cheap oil is over" (IEA 2008, 3). Because oil is the lifeblood of industrial society, and because it is so difficult to find adequate substitutes, oil scarcity may have momentous consequences for the future world economic outlook (IMF 2011, 89–124).

Fueling the Future

What does all of this mean for the future world energy mix? The answer depends on whether one focuses on energy scarcity or climate change, or both. Either way there is a need to move away from fossil fuels, but not all fossil fuels are created equal. From an energy scarcity viewpoint, some fossil fuels are scarcer than others, and for the time being only oil is becoming seriously scarce. From a climate change viewpoint, some fossil fuels are more damaging than others: coal is higher in carbon than oil, which in turn is higher in carbon than gas.

From the viewpoint of climate change, the best thing that could ever happen to the planet is to run out of combustibles, and the fact that coal remains in abundant supply is bad news. But it is rather good news from the viewpoint of oil depletion because switching the energy mix toward other fossil fuels, such as coal and gas, is much cheaper than toward renewable energies, such as wind and solar.

The only responsible approach is of course to take both climate change and fuel depletion into account. From such a perspective, a transition away from fossil fuels to renewable forms of energy seems like an obvious no-regret strategy. But this is more easily said than done. As we will see, a balanced account of future fuel options must not only take both climate change and oil scarcity into account but also consider fundamental practical issues of technological feasibility and commercial attractiveness.

Carbon Footprints

For starters, let us note the different carbon footprint of different fuels. A useful measure is the amount of CO_2 emitted per kilowatt hour (kWh) of electricity produced. Coal has the highest carbon footprint, ranging from 830 to 940 grams of CO_2 per kWh or even higher, depending on the sort of coal. Although in most countries oil has become too precious to be wasted on electricity generation, the production of 1 kWh from oil leads to "only" 610 grams of CO_2 emissions. This is an indicator of oil's somewhat lower carbon footprint. Natural gas is even less damaging to the climate, with the production of 1 kWh from gas resulting in emissions of only 370 grams of CO_2 (IEA 2011b, 39).

Nuclear power generation is completely carbon free, but considerable amounts of CO_2 are released when mining uranium, building nuclear stations, and so on. The same applies to renewable energy. A wind park or solar facility operates without emitting any carbon, but emissions do occur at other stages such as construction or decommissioning. In most cases, the carbon footprint of renewable energy is considerably lower than that of any fossil fuel. Sometimes, however, the embedded amount of carbon can be considerable. This happens, for example, when biomass is produced from industrial agriculture with heavy use of petrochemical products and fossil-fuelled machinery, despite the fact that, in theory, the burning of biomass releases to the atmosphere exactly the same amount of carbon that was previously withdrawn during plant growth.

Stuck with Oil

As the history of energy shows, replacing one energy source with another is never fast and easy (Yergin 2011). This also applies to the substitution of oil in the present historical juncture (IMF 2011, 89–124). The main reason why oil is so hard to replace is its key role as a liquid fuel. According to the IEA (2012a, 87–97, 552), the transport sector is 93 percent oil-based and represents well over 50 percent of world oil consumption. Even by 2035, the contribution of biofuels and natural gas to the transport sector is expected to be no more than 11 percent. The expectation for electricity is even more disappointing, with a meager 2 percent. Most electric transport will happen in the railway sector whereas, despite all the hype, electric cars (including hybrids) are expected to make a fairly limited contribution to world road transport until 2035.

To some extent gas may replace oil as a transportation fuel because buses and other vehicles can be retrofitted to run on gas. The development of an appropriate refueling infrastructure, however, remains a challenge. The same applies to electricity and, even more, to technological dreams about hydrogen.

In principle, there are various technical procedures to convert coal and gas into liquid fuels. But it is important to note that the liquefaction of coal and gas has extremely serious downsides. It is not only very expensive but also extremely carbon intensive, and the final product comes with a fairly low EROI.

Unconventional Oil

To some extent, declining production of conventional crude can be offset by fuel produced from unconventional sources such as extra-heavy oil, bitumen recovered from tar sands, or shale oil. It is important to bear in mind, however, that the high production cost of unconventional oil unavoidably translates into a higher price. Also, unconventional oil has a higher carbon footprint than conventional crude (Brandt and Farrell 2007). Moreover, it has a low EROI due to the fact that considerable energy inputs are needed for its production (Verbruggen and Al Marchohi 2010). EROI is further reduced when technologies are deployed to mitigate the negative climatic and environmental side effects.

Regardless of the harmful consequences, it is reasonable to expect that unconventional oil will be increasingly produced. By the same token, highly challenging oil reserves from the Arctic to Antarctica are likely to be cleared for exploitation.

Natural Gas

Natural gas is one of the most appealing energy resources to mitigate both climate change and oil scarcity. While gas reserves are limited, it is far more abundant than oil and its carbon footprint is lower than that of any other fossil fuel. Nevertheless, even under the best of circumstances gas can only be a transition fuel. "Although gas is the cleanest of the fossil fuels, increased use of gas in itself . . . will not be enough to put us on a carbon emissions path consistent with limiting the rise in average global temperatures to 2°C" (IEA 2011a, 155). Also, gas poses considerable commercial and practical problems. While oil and coal are easily transportable, gas requires pipelines or needs to be liquefied.

Recently there are encouraging developments in the extraction of unconventional gas, most notably shale gas. However this so-called shale gas revolution has been limited to the United States, and may remain so for a while due to serious environmental concerns, as well as legal restrictions and logistic constraints (Stevens 2010; IEA 2012c). Also, it is important to note that the carbon footprint for shale gas is higher than for conventional gas.

King Coal

World coal reserves remain abundant, contrary to alarmist views about a near or imminent peak of coal production (Energy Watch Group 2007; Heinberg 2009a; Höök et al. 2010). From 1990 to 2010, the share of coal in the world energy mix has risen from 25 to 27 percent, and it is still increasing. Especially in China and India, there has been a huge surge in coal consumption. The resource base of coal is vast and geographically diverse, with enormous deposits in the United States, China, India, Russia, and Australia (IEA 2012a, 155–177, 552; see also Höök and Aleklett 2009; Lin and Liu 2010; Lin et al. 2012).[15]

From an energy scarcity viewpoint, coal remains an obvious fallback position for decades. Coal was the basis of industrial society before the start of the oil age, and it may regain some of its erstwhile importance after peak oil. Despite the fact that high oil prices are making coal mining and transportation more expensive, producers in coal-rich countries will be motivated to tackle such challenges as long as coal production makes sense in terms of financial and energy returns.

From a climate change viewpoint, however, coal has the highest carbon footprint of all fossil fuels. Therefore, its share in the world energy mix needs to decline drastically rather than increase further. In fact, the harmful climatic and environmental consequences of a few more decades of constant or even rising carbon emissions from coal are predictable: unless we assume the adoption of appropriate clean coal technologies, most notably carbon capture and storage, catastrophic global warming far above 2°C will become unavoidable.

In principle, CCS has significant potential to reduce the carbon intensity of coal and, to a limited extent, also gas used in power and large industrial plants. It is necessary, however, to see the uncertainties and limitations of this technology. First, CCS is a long way from being available on an industrial scale and may not become so early enough to avert runaway climate change. Second, it is applicable only in large-scale industrial operations such as power generation, not in transport and other small-scale activities. Third, like most other clean energy technologies, the application of CCS requires significant investment and reduces EROI.

Nuclear Energy

Nuclear power is sometimes contemplated as a silver bullet to "fix" peak oil and climate change with one stroke. Indeed, the carbon footprint of nuclear power is low. There are certain carbon emissions along the nuclear lifecycle, for example when mining uranium, but even so the footprint is much lower than for fossil fuels (Sovacool 2008; Warner and Heath 2012). Moreover, uranium reserves are fairly abundant though of course not unlimited (Moriarty and Honnery 2009, 2472). The lifetime of the world's uranium reserves can be considerably extended if fast-breeder technology is introduced to reprocess spent fuel rods.[16]

Let us not forget, however, that the current share of nuclear in the world energy mix is less than 6 percent. Nuclear power plants have very long lead times, so significantly increasing that share will take time. In fact, the IEA does not expect nuclear to rise to more than 7 percent by 2035 (IEA 2012a, 552). The reasons are well known. Nuclear power generation is considered to involve incalculable risks, including nuclear proliferation. If we assume a crisis, then investment may dry up and planning horizons are likely to shrink below the fifteen years that are required as a minimum to build a nuclear facility.

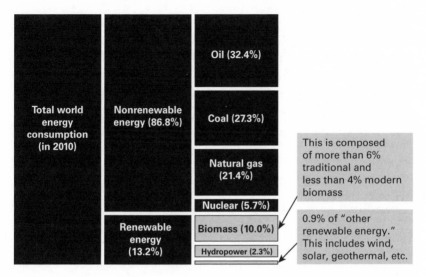

Figure 2.6
The world energy mix in 2010. *Source:* IEA 2012e, 6.

Renewable Energy

Like nuclear power, the importance and potential of renewable energy must be assessed in relation to the global energy mix, which is visualized by figure 2.6. As the figure indicates, renewable energy accounted for 13.2 percent of the world's total primary energy consumption in 2010.

This number becomes considerably less impressive when we contemplate that the largest share of renewable energy is biomass, which encompasses both what is called modern and what is called traditional biomass. For modern biomass, think of wood pellets or biofuels. For traditional biomass, think of Ethiopian women collecting firewood or Mongolian herdsmen burning camel dung. The second largest share of renewable energy is hydropower, which includes gigantic and environmentally damaging installations such as China's Three Gorges Dam.

The genuinely green renewable energy, classified as "other renewable energy" in figure 2.6, makes up no more than a meager 0.9 percent of the world energy mix. This includes anything from wind to solar and from geothermal to tidal power. Many commentators set their hope on this tiny fraction. The obvious problem with such wishful thinking is that, despite the immense cost, even multiplying this 0.9 percent by an order of magnitude will not solve either climate change or energy scarcity.

From a less green viewpoint, modern renewable energy can be understood as also including modern biomass and hydropower, despite the dubious environmental pedigree of the latter. But even if we do that, we still have to exclude more than 6 percent of traditional biomass. Subtracting traditional biomass, the share of modern renewable energy constitutes no more than 7 percent of the world energy mix (IEA 2012a, 214, 552). This is slightly better than nuclear energy's share, but it is nowhere near the importance of any fossil fuel such as oil, coal, or gas.

Under its New Policies Scenario, which is consistent with global warming of 3.6°C, the IEA projects the share of renewable energy to increase from 13 to 18 percent by 2035. The share of *modern* renewable energy, as defined above, is expected to double from 7 to 14 percent (IEA 2012a, 214, 552). This means that, despite very high growth rates, even in 2035 the share of renewables in the world energy mix is likely to remain significantly lower than that of any fossil fuel (figure 2.7).

Based on historical experience, it would be unreasonable to expect more rapid growth of renewable energy (Smil 2010; Haberl et al. 2011; Höök et al. 2012; Fouquet and Pearson 2012; Allen 2012; Pearson and Foxon 2012). Even the limited growth expressed in figure 2.7 rests on the assumption that the world economic and political situation will

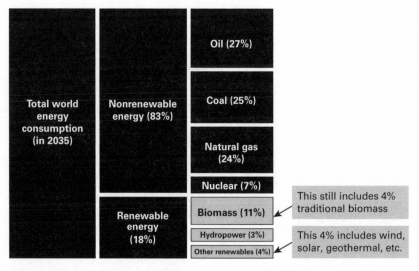

Figure 2.7
The world energy mix in 2035. *Source:* IEA 2012a, 552.

allow for ambitious energy programs. Renewable energy requires considerable investment, with the IEA stating that "Renewable energy subsidies jumped to $88 billion in 2011, 24% higher than in 2010, and need to rise to almost $240 billion in 2035 to achieve the trends projected in the New Policies Scenario" (IEA 2012a, 211; note that the value for 2035 is net of expected inflation). Under crisis conditions, this may not be forthcoming. Conventional fossil fuels such as coal and gas are economically more attractive and more easily available than renewable energy.

Another crux is that the most environmentally appealing forms of renewable energy such as wind and solar have the serious downside of intermittency (Trainer 2012). For example, an offshore wind park typically requires a non-renewable power plant as a backup for when the sea breeze is not blowing. The problem can be solved with an expensive connected infrastructure, or super grid, but that requires truly significant investment in terms of finance and public subsidies.

Sure enough, once an offshore wind park and related infrastructure are up and running, they require no energy input and are entirely carbon free. But their construction requires considerable energy, pushing down the overall EROI, as well as raw materials such as concrete and steel, embodying significant CO_2 emissions. Even more than wind and solar, biofuels have a disappointingly low EROI because they require considerable energy inputs from industrial agriculture (Gupta and Hall 2011). At least so far, they have a tendency to displace food production and to crowd out important ecosystem services. When a forest is cleared to enable biofuels production, the carbon balance is appalling due to the release of the carbon that was previously stored in the forest.

The crunch is that solar cells do not produce solar cells, and wind turbines do not produce wind turbines. So far, most of the energy inputs necessary to produce renewable energy have been supplied from non-renewable sources. In terms of energy flows, one may therefore say that non-renewable energy is subsidizing renewable energy. Expanding the share of renewables beyond 20–30 percent is very challenging because, at some point, the production and maintenance of the renewable energy infrastructure will have to be fueled by renewable energy. If renewables are to become the main source of energy, then they must cater to the reproduction of their own energy infrastructure. This is a tall order, and the global potential for renewable energy is limited (Moriarty and Honnery 2012).

At best, the renewables revolution is an enormously demanding task, insofar as it requires a radical shift from incremental to systemic change. At worst it is impossible for renewables to sustain a consumer society (Trainer 2007).

The Energy Quagmire

For generations, the energy backbone of industrial society has been oil. Now peak oilers predict a rapidly accelerating decline of world oil production. If they are right, oil will not be able to meet the projected needs of the world economy. Gas may eventually provide a temporary substitute, but it is far from carbon free and it is not a liquid fuel. The other available surrogates are even less appealing. Coal is abundant and cheap, but it is associated with very significant carbon emissions and like gas it is not a liquid fuel. Nuclear is largely carbon free, but uranium is not abundant—in addition, it is not cheap and there are notorious security risks emanating from nuclear power plants. Renewable energy is also not cheap. While nuclear and renewable energy are most attractive from the viewpoint of climate change, in both cases there are infrastructural constraints on how far and how quickly their share in the energy mix can be expanded.

Any solution is likely to be based on a combination of four elements: lower energy consumption, better energy efficiency, a switch from fossil fuels to other forms of energy, and carbon capture and storage. A forced curtailment of energy consumption would be enormously painful, so the million dollar question appears to be whether it can be avoided by a combination of the other three.

Bold Visions
The most radical voices in the peak oil scene claim that any surrogates for cheap oil are incompatible with high rates of economic growth (Heinberg 2007, 2009b, 2011). Similarly, the most radical voices in the climate change community claim that anything short of a radical transformation of industrial society leaves us on a path to climate disaster. To prove such "doomsayers" wrong, what would be needed is a radical crash program

to develop and implement a mix of surrogate resources and adequate new technologies, combined with drastically improved energy efficiency. The program would have to start early, as it would take well over a decade to come to fruition (Hirsch, Bezdek, and Wendling 2005, 2010).

Bold visions can be found in the *Energy Technology Perspectives*, a biennial publication by the International Energy Agency. In the BLUE Map (2010b) and in the 2°C Scenario (2012b), the IEA forecasts that baseline CO_2 emissions can be cut by more than half by 2050, and further down to zero by 2075 (IEA 2012b, 513–533).[17] The IEA makes heroic assumptions on improved energy efficiency and reduced carbon intensity, suggesting that all good things can go together. Not only can primary energy consumption rise by a further 32 percent, but payback from fuel savings will make this low-carbon or even zero-carbon economy possible on a net gain equivalent to $60 trillion of additional world GDP (IEA 2012b, 38).

Undoubtedly, such a "truly global and integrated energy technology revolution" (IEA 2010b, 60) presupposes processes far more optimal than anything ever seen in history. The correlation between economic activity, energy consumption, and carbon emissions must be broken "through a transformation of the global energy system and its technologies" (IEA 2012b, 30). To the extent that there are obvious limitations to decoupling, the energy system must be decarbonized. This includes a huge increase of biofuels production, which together with food production would occupy most of the planet's bioproductivity. Humanity would capture the last remnants of what used to be called nature. As the IEA notes, the exercise is not difficult "from a modeling perspective" (IEA 2012b, 517). But how would a planet almost completely dedicated to biofuels and food production look like?

Fortunately or unfortunately depending on how you see it, the energy revolution envisaged by the IEA will not come to pass.[18] Under current policies, both energy consumption and carbon emissions would double by 2050 rather than being cut by 75 percent. To put it mildly: "While efficiency measures have achieved some reduction in global energy intensity, the rate of improvement has slowed in recent years, which is worrisome" (IEA 2012b, 30).

Another challenge with increased energy efficiency is that adequate measures would have to be taken to prevent it from being frustrated by the so-called rebound effect. The rebound effect, or Jevons paradox, denotes the risk that better efficiency may not lead to lower consumption because it reduces cost, which, in turn, encourages higher consumption (Sorrell 2007; Polimeni et al. 2008)[19].

From a technical viewpoint, it would be appealing to rely more on electricity to reduce the dependency of industrial society on any particular fuel, such as gasoline. Anything can be converted to power, and once it is power the source does not matter. In practice, however, the conversion of the existing vehicle fleet from gasoline to electricity is a tall order. As we have already seen, even the IEA does not expect electric cars to play a significant role before 2035 (IEA 2012a, 91).

And even if we assume that further electrification is possible at affordable cost and within a reasonable timeframe, it is not *per se* a solution to climate change. It will not reduce our dependency on fossil fuels, unless the additional electricity is produced from nuclear power and/or renewable energy, or, as a second-best option, if the carbon emissions are captured and stored. Alas, we have already seen why a global renewables revolution and a rapid surge of CCS are unlikely.

Like it or not, the IEA's infamous 6°C Scenario remains the trajectory we are on. When you open the news, do you see any serious crash program underway or in the offing? Once the post–peak energy crunch has seriously started, rather than grandiose designs, expect haphazard moves to handle a difficult situation.[20]

Catch-22

In thermodynamic terms, industrial society is threatened by two bottlenecks: at source by the limited availability of oil, and at sink by the limited capacity of the atmosphere to absorb CO_2 emissions without wreaking havoc on natural ecosystems and human societies. Fossil fuels are at the root of the problem in either case, but for opposite reasons. In the case of climate change, the *general* abundance of fossil fuels is a significant part of the problem. In the case of peak oil, the problem is the scarcity of a *particular* but vitally important fuel.

There is a comfortable but naive way to see the connections between climate change and peak oil, and there is a more realistic but less comfortable perspective. The starry-eyed view is that both climate change and peak oil mandate a radical reduction of fossil fuel consumption, with

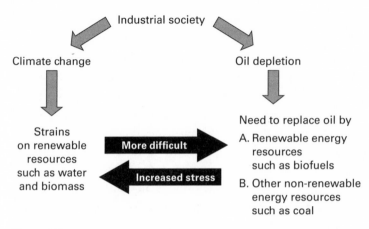

Figure 2.8
Tradeoff between the mitigation of climate change and peak oil

one challenge providing an additional incentive to deal with the other. From such an ideal perspective, a shift to renewable energy is an obvious no-regrets strategy to deal with both peak oil and climate change (Rozenberg et al. 2010).

Unfortunately, the real world is a far cry from win-win. As figure 2.8 suggests, the constellation of climate change and peak oil produces an intensely dilemmatic situation (Bradshaw 2010). Climate change leads to strains on renewable resources such as water and biomass. For example, if rainfall patterns become unpredictable it will not only become a daunting challenge to feed the world's growing population but it will also become more difficult to replace oil by renewable energy such as hydropower and biofuels. But if oil is replaced by fossil fuels such as coal or gas, then this increases the stress on the climate system.

Humanity is thus caught in a catch-22 situation. On the one hand, the most readily available substitutes for oil, such as coal or tar sands, are even more carbon-intensive than crude. As this would exacerbate climate change, it would be much better to substitute oil with renewable resources. On the other hand, many renewable resources such as water and biomass will be under increasing strain from climate change. At the same time, the prospect of changing meteorological and hydrological patterns will make it more difficult to plan a renewable energy infrastructure based on wind and hydropower. Food production is also going to become more difficult for the same reasons, which will then make it even more difficult to rededicate land for the production of biofuels.

Another serious problem is that fossil fuels and other non-renewable resources are increasingly needed to reduce strains on renewable resources. This in turn speeds up climate change and may worsen resource shortages further down the line. For example, the industrial production of food and energy crops heavily relies on energy and chemical inputs based on fossil fuels; thus, industrial agriculture accelerates both climate change and resource depletion. Another case in point is desalinated sea water, the production of which consumes an enormous amount of energy that is typically derived from fossil fuels. Sometimes, irrigation and drinking water can also be extracted from geological aquifers (in the south-western United States, for example), but what will happen when these aquifers run dry? While all of this temporarily mitigates strains on renewable resources, it aggravates climate change and future resource scarcities.

In times of climate change, renewable resources such as water and biomass may be hard pressed even to safeguard human subsistence, let alone solve the energy problem.

Bottom Line

Taken together, abrupt climate change and disruptive energy scarcity have the potential to plunge industrial civilization into agony. Not only are they closely intertwined, but various assumptions can be made about their sequencing. The worst-case scenario is a pincer movement with peak oil first causing a shift to more carbon-intensive technologies, thus accelerating anthropogenic climate change and further reducing the long-term carrying capacity of the planet; and with runaway climate change then mandating a radical curtailment of fossil fuel consumption or simply delivering the coup de grace to industrial civilization.

3

What the Climate Can Change

Society can change climate, and climate can change society. Climate change can have social and political effects in many different ways, but the most basic ones are related to human needs. Our subsistence depends on food, drink, and shelter. Food depends on agriculture, while drink depends on fresh water. Agriculture also depends on fresh water, as well as fertile land. Unfortunately, climate change is expected to have dramatic consequences for the availability of fresh water and fertile land; and, thereby, for access to food and drink. The third basic human need, shelter, requires a stable land base. Alas, climate change is expected to affect that land base via rising sea levels and other losses of human habitat.

Even without climate change, feeding and housing nine billion people by 2050 is a big challenge. Add climate change, and the task becomes daunting. Consider that, in combination with energy scarcity, climate change can throw industrial civilization into agony, and you get a real sense of the magnitude of the challenge. Obviously, this does not mean that we will all die from hunger or thirst as drink and food become scarce, or that we will all drown as sea levels rise. Clearly, the social and political effects of climate change are more complex than that, and they vary by context. As this chapter shows, they are mediated by a multitude of social processes ranging from conflict and migration to large-scale systemic crisis.

The work of the Intergovernmental Panel on Climate Change (IPCC) is an obvious starting point. Unfortunately, the IPCC provides only few clues about the social and political effects of climate change, so I move on to examine the work of eco-scarcity theorists who have extensively discussed the social and political effects of strains on environmental resources such as water and arable land. This is relevant because, in the event of climate change and possibly in combination with peak oil, a declining industrial civilization may lose its erstwhile ability to mitigate the negative effects of strains on environmental resources.

To understand what might happen under such circumstances, it is most useful to look at the effects of climate change episodes *before* the advent of industrial civilization. As a baseline, I present a model that was originally developed to account for the social and political dislocations caused by climatic cooling in the early modern period (Zhang et al. 2007; 2011). I argue that, paradoxically, the social and political effects of climatic cooling are similar in kind to the effects of climatic warming because, regardless of the nature of the forcing, the decisive issue at stake is whether or not human society is able to cope with that forcing.

The challenge becomes obvious if we take a complexity perspective. Societies have a clear tendency to accrue organizational complexity and technological sophistication. While this complexity is hard to pin down because it can take so many forms, it enables societies to solve a variety of problems. At the same time, complexity also creates its own problems. The net effect of these two tendencies is that over time the returns on investment in further complexity diminish. Once the returns are very low or even negative, the next serious problem to emerge leads to an existential crisis in which the preservation of a society hinges on its ability to revolutionize its problem-solving capacity by transitioning to a new regime of higher organizational and technological complexity. Tragically, such a transition is not always viable. Societal collapse, understood as the forced reduction of complexity resulting from an existential crisis, is another possibility (Tainter 1988).

My first set of cases is drawn from the ancient Near East, 11,000–1000 BC. During these ten millennia, Mesopotamia and its environs were challenged by four significant climate change episodes. While the responses to the first couple of episodes were largely progressive and led to civilizational upgrades, in the third and fourth episodes most of the Near East's highly advanced civilizations were overwhelmed. This suggests that progressive adaptations to climate change are more likely at the early stages of a civilization when there are large unexploited gains from investment in higher complexity. As the marginal cost of higher complexity increases, similar upgrades become more difficult and, at some point, impossible.

My second set of cases is about the medieval Norse settlements on Iceland and Greenland, 900–1500 AD. In contrast to the advanced civilizations of the ancient Near East, the focus here is on marginal human habitats. The settlement in Iceland was flexible enough to weather the transition from the Medieval Warm Period to the Little Ice Age, whereas the settlement in Greenland was less adaptable and collapsed. This suggests that the survival of marginal communities largely depends on their cultural flexibility, and thus their resilience to changing environmental circumstances.

When applying this to the future, I find that industrial society is at a level of complexity where it is increasingly hard to envisage progressive and complex solutions to cope with severe stresses such as climate change. At the same time, a *voluntary* return to sustainability at a lower level of social and political complexity is unlikely because society has invested in all its complexity for a reason, namely to solve real problems. This is of course not to deny that, subsequent to the collapse of industrial society, the fragments are likely to reassemble themselves into lower-complexity social and political systems.

Contrary to the conventional view that people from poor countries are most vulnerable to climate change, I argue that in many cases they are better placed to recover from the consequences of systemic breakdown. One reason is that, more often than not, their social cohesiveness and thus their community resilience are better than for people from rich industrialized countries. Moreover, as my historical case studies from medieval Greenland and Iceland suggest, the more flexible and adaptive a community, the more successfully it is likely to weather the significant environmental stresses emanating from climate change.

The IPCC Approach

When looking for insight about the social and political effects of climate change, the most obvious place to start is Working Group II of the Intergovernmental Panel on Climate Change. The IPCC is composed of three working groups. The mandate of Working Group II is to deal with "impacts, adaptation and vulnerability."[1] Given this mandate, one would expect relevant insights about the social and political effects of climate change from Working Group II. Most notably, one would expect to find such insights in the weighty tomes the group has contributed to recent assessment reports (IPCC 2001c, 2007b).

In the latest report of Working Group II (IPCC 2007c), most attention is dedicated to the direct impacts of climate change. This includes the availability of livelihoods such as water and food, as well as the degradation of ecosystems. There is also considerable attention to the loss and degradation of land, both as human habitat and for agriculture. The worst impacts are expected in low-lying and coastal areas, due to flood risks related to sea level rise and extreme weather. There is also some attention to the health effects of climate change on people.[2]

While Working Group II concentrates on the direct impacts of climate change on natural, physical, and biological systems, it hardly covers the indirect effects on social and political systems. This is not to deny that, in a few places, the group makes timid attempts to cover the social effects. For example, there is some talk about "industry, settlement and society." There is also some attention to the links between climate change and sustainable socioeconomic development. Presumably under the influence of Nicholas Stern (2007), this includes the overall economic effects of climate change. Despite such digressions, there is scant attention to the more genuinely social and political effects. To cite only the most glaring example, it can be shown that the IPCC has given highly summary and neglectful treatment to the academic discussion about a possible nexus between climate change and violent conflict (Nordås and Gleditsch 2009).

This reticence is understandable. There is far more certainty about the direct impacts of climate change on natural, physical, and biological systems than about the indirect effects on social and political systems. The IPCC must be careful to avoid any controversial claims lest it be mercilessly criticized by climate skeptics.

Nevertheless, the scant attention the IPCC has dedicated to the social and political effects of climate change is disappointing because some of the most momentous changes for the people affected by climate change will be social and political in nature, rather than merely physical or physiological. It is also disappointing because the IPCC is officially inspired by a human-security perspective (Detraz and Betsill 2009). Given this official focus, the IPCC can hardly justify its neglect of the most fundamental social and political effects of climate change.[3]

What we can gain from the IPCC is a suitable starting point for our discussion. Strains on environmental resources such as fresh water and food are likely to trigger the most significant social and political effects of climate change (Godfray et al. 2010). The loss and degradation of land, most notably in coastal plains, is also crucial because it may displace large populations (K. Smith 2011). As we will see in the next section, the literature on eco-scarcity can tell us more about the likely consequences of these stresses for social and political order.

Eco-Scarcity

Thomas Malthus (1798) remains the *locus classicus* for anybody interested in the social and political consequences of scarcities resulting from strains on renewable resources. His original theory accounts for preindustrial cycles of overpopulation outpacing the increase of agricultural productivity, and thus leading to severe social and political troubles or even large-scale systems breakdown.

Despite the brilliance of his theory, Malthus turned out to be wrong due to his failure to anticipate the Industrial Revolution. As it happens, the Industrial Revolution and the concomitant intensification of trade has made it possible for agricultural productivity not only to keep pace with population growth, but also to sustain increasing economic wealth for the majority of people in advanced industrial countries (see the related discussion in chapter 1).

Due to its axiomatic nature, Malthusianism has never lost its grip on radical thinkers. As with any elegant theory, its adepts have proposed various updates to demonstrate its continued validity in the face of countervailing evidence. A recent example of this is eco-scarcity theory, whereby land degradation and other environmental strains combine with population pressure to unleash Malthusian scenarios of social conflict and political disorder.

Eco-scarcity theory began in the 1990s with conflict theorists suggesting complex causal links between environmental pressure, defined as scarcities of renewable resources, and the outbreak of violent conflict.[4]

Their strategy was to collect case studies substantiating the claim that, particularly in overpopulated developing countries, environmental pressure can lead to the outbreak of violence. Two ample collections of case studies were produced roughly at the same time, one by a Canadian team (Homer-Dixon 1994, 1999) and the other by a team based in Switzerland (Bächler et al. 1996). Both of these teams focused on cases of violent conflict in developing countries, and both had the aim of tracing the social processes leading from environmental scarcity, eventually combined with population pressure, to the outbreak of violent conflict.

The exclusive reliance on cases where environmental pressure was associated with violent conflict had one great problem and one great advantage. The problem was a serious confirmation bias: all case studies were selected on the basis of what was to be explained and what was supposed to do the explaining. In other words, the case studies were selected on the following two requirements: (1) that violent conflict actually did break out, and (2) that some form of environmental pressure preceded it. As a consequence of this practice, researchers were almost certain to find what they were looking for.

Despite this serious bias, the procedure chosen by early eco-scarcity theorists had an important advantage: it placed the spotlight on the social processes taking place in those cases where violent conflicts had in fact been driven by environmental pressure. Thomas Homer-Dixon (1994, 31), the leader of the Canadian team, presented these "mechanisms" in a neat causal model.[5]

According to this model (figure 3.1), environmental scarcity is triggered by a combination of population growth and excessive strain on some vital but dwindling renewable resource, typically exacerbated by unequal access to that resource. Together with the direct effects of the scarcity itself, the ensuing economic crisis engenders the forcible displacement of people and/or their voluntary emigration. The result is social segregation and a weakening of state structures, both in the country affected by environmental scarcity and in neighboring countries targeted by a massive inflow of migrants. In some cases this may lead to a coup d'état or even state breakdown.

There are two ways in which all of this increases the risk of violent conflict. First, scarcity-driven migration may provoke violent clashes between the migrant population displaced by environmental pressure

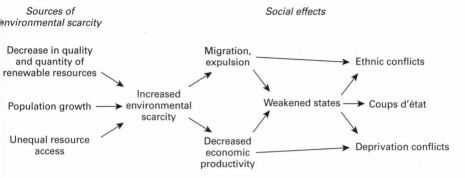

Figure 3.1
Causal pathways from environmental scarcity to violent conflict

and the recipient population (ethnic conflicts). Second, the economic crisis in the area of origin, combined with a declining ability of the state to manage that crisis, can lead to an insurgency of citizens who feel deprived of the standard of living they either feel entitled to, or need in order to survive (deprivation conflicts).

Neo-Malthusian models are easy targets for empirical criticism because, like Thomas Malthus's original theory, they fail to take into account the systemic effects of modernity. They are logical deductions from a theoretical model that, at least sometimes and in some places, applied before the advent of industrial civilization (LeBlanc 2003); would apply in the absence of industrial civilization; and will again apply after its terminal demise. So long as industrial society remains triumphant, however, it is easy to come up with countervailing qualitative case studies to "falsify" eco-scarcity theory (e.g., Peluso and Watts 2001). Similarly, it is hardly surprising that the balance of recent quantitative studies examining conflict datasets do not support the claim that environmental pressures have any statistically significant causal effect on violent conflict (see Bernauer, Böhmelt, and Koubi 2012).[6]

To be sure, the quantitative literature debunking eco-scarcity can itself be criticized. This has sometimes been recognized even by quantitative scholars: "Conventional statistical techniques run into problems when the relationships to be investigated are of a complex and interactive kind, which is exactly the case for eco-scarcity theory" (Theisen 2008, 814). Moreover, statistical analysis is methodologically skewed toward proximate rather than remote causes. For example, mortality cannot be proven to be the cause of death by correlation analysis. After all, people die from diseases and accidents but not from mortality. And yet we all know that mortality is why we must die. Similarly, conflicts may be caused by environmental pressure even if the triggers are the usual suspects such as ethnic hatred or social injustice.

Another shortcoming of the quantitative literature criticizing eco-scarcity is its narrow focus on violent conflict. The original model by Homer-Dixon reproduced in figure 3.1 suggests that, in addition to violent conflict, environmental pressure can have various other consequences: migration, state destabilization, and coups d'état. It is highly unfortunate, to say the least, that violent conflict has come to be seen in isolation from such broader social and political consequences of environmental pressure.

This is not to deny that, when measured against its own claims, eco-scarcity theory is in trouble if there is no demonstrable statistical correlation linking environmental pressure with violent conflict. The absence of a clear statistical link does undermine the applicability of this neo-Malthusian school of thought for the analysis of conflict patterns in the recent past and in the present. But it does not alter the fact that Malthusian scenarios may well be borne out in the future if industrial civilization enters a terminal decline.

The Climate Watershed

Ever since the Industrial Revolution, the systemic effects of industrial civilization have belied Malthusian expectations. For example, contrary to the predictions of Thomas Malthus, industrial agriculture has for a long time enabled exponential population growth. This and similar effects operate while industrialized society lasts, but will be overridden by the demise of industrial civilization. If climate change makes industrial society unviable, eco-scarcity and other Malthusianism scenarios are bound to return with a vengeance.

Today, industrial civilization is buttressing a globalized world system that injects trade, aid, know-how, and governance capabilities to some of the most vulnerable parts of the world, which would otherwise suffer severe social and political dislocations from environmental stresses. In our globalized world, even the poorest and most vulnerable countries and their societies are embedded in industrial civilization, both by virtue of transnational interdependence and through governmental links such as development aid and military intervention. In a way, the industrial era with its enormous energy inputs and technological inventiveness has created a fool's paradise that temporarily abrogates the worst effects of environmental pressure.

This does not always apply to the extent desirable from a humanitarian viewpoint, and eco-scarcity theorists eagerly pick on cases where the so-called international community has failed to prevent or mitigate violent conflict caused by environmental pressure. Nevertheless, in most places most of the time, the violent consequences of environmental pressure have been blocked by world industrial civilization. While environmental stresses may have led to violent conflict in some remote parts of the developing world, everywhere else they have increased the incentive for international cooperation (Dinar 2011). But alas, this applies only as long as world industrial civilization is in a position to bail out places struck by environmental pressure. Once it enters a terminal decline, whether due to climate change or any other reason such as energy scarcity, Malthusian fears may yet turn out vindicated.

In a significantly hotter world suffering from increasing energy scarcity, either due to peak oil or because rationing the consumption of fossil fuel turns out to be the only practical way to prevent runaway climate change, we cannot simply presume that industrial civilization is going to remain steadfast.

This is not to deny that, at the initial stages of the crisis, industrial civilization can easily mitigate some of the social and political effects of climate change. Later on, however, as the crisis becomes more serious, world industrial civilization may lose its erstwhile ability to bail out places in mayhem. International cooperation will then become more intermittent, and far more people are likely to be let down by the international community than have been in recent decades. At some point, environmental pressures may lead to serious problems even in places that are firmly in the industrial core.

In other words, the near future may be radically different from the recent past. Contrary to the bromide that nature does not make jumps (*natura non facit saltūs*), climate change not only leads to a gradual increase in temperature that slowly and continuously moves an envelope of variability in one direction, but also to an increased likelihood of extreme weather events (see chapter 2). Moreover, there is a risk that radical discontinuities due to feedback effects and tipping points may lead to largely unpredictable systemic consequences (Lenton et al. 2008). From this perspective, the gradualist view of continuous warming conveyed by most climate models is dangerously misleading. Or, as Milly and colleagues (2008) have put it, "Stationarity is dead."

Just as neo-Malthusian proponents of eco-scarcity theory fail to acknowledge that we are still living in the industrial age, their cornucopian critics fail to appreciate that the durability of industrial society cannot be taken for granted in a turbulent world. Climate change and other global risks such as peak oil are game changers that may drive the world toward a post-industrial and post-global age where we may see precisely the neo-Malthusian scenarios that are so often discarded.

Climate-based neo-Malthusianism started with causal models similar to eco-scarcity theory, often highlighting environmental migration as an important factor intervening between climate change and violent conflict (Barnett and Adger 2007; Reuveny 2007). As in the case of eco-scarcity theory, this was countered by arguments based on the statistical analysis of recent events, typically highlighting the absence of a statistically significant causal link connecting climate change with the outbreak of violent conflict (e.g., Raleigh and Urdal 2007).

Remarkably, however, even authors representing the variable-based approach sometimes acknowledge that statistical models based on recent historical events are unable to predict the conflict dynamic under abrupt climate change: "We are only beginning to experience the physical changes imposed by global warming . . . so a lack of systematic association between the environment and armed conflict today need not imply that such a connection cannot materialize tomorrow" (Buhaug, Gleditsch, and Theisen 2010, 93–94).

In any case, the causal mechanisms under scrutiny are strikingly similar to those previously developed by eco-scarcity theorists. For example, consider the model in figure 3.2 (Buhaug, Gleditsch, and Theisen 2010, 82).

The model is remarkably sophisticated, belying allegations of "environmental determinism."[7] However, as in the case of the eco-scarcity

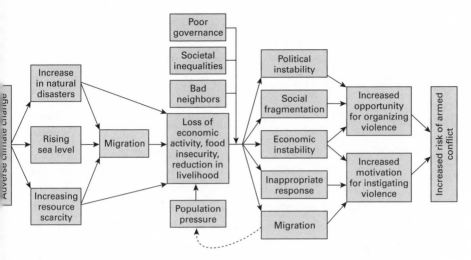

Figure 3.2
Causal pathways from climate change to armed conflict

literature, it is important not to forget that violent conflict is only one of many possible social and political effects of environmental pressure—alongside state failure, coups d'état, and peaceful forms of environmental migration.

Back to the Future

Sometimes the recent past is a poor guide to the future, just as what you see in the rear-view mirror of your car is a poor indicator of the road lying ahead. To overcome the comfortable but deceptive cognitive habit of assuming that tomorrow will always be like yesterday, it is helpful to look at more distant historical periods and episodes for insight on how the social and political consequences of environmental pressures may unfold.

This is not to deny the inherent difference between pre-industrial societies and the situation to be expected when industrial civilization enters an existential crisis. It is obvious that the technological capacities of industrialism will not disappear overnight. But when it comes to environmental scarcities, these capacities may be as much part of the problem as part of the solution. For example, military capacities can be utilized as readily for warfare as for peacekeeping. Thus, the positive and negative effects of industrial civilization may easily cancel each other out. Even so, and bearing the differences in mind, the best available proxy for the way societies are likely to react to climate change once industrialism enters a decline is the way societies have reacted to comparable crises before the consolidation of industrialism.

In two separate studies presented four years apart, David Zhang and his colleagues looked at the period between 1500 and 1800 to understand the social and political effects of climate change (Zhang et al. 2007; 2011). They have used time series from the Northern Hemisphere, especially from Europe and to a lesser extent from China, to develop and refine their theory. Despite the obvious limitations to the quality of the data available, and despite the unavoidable lack of nuance in their macro-historical exercise, in their second publication, the authors (Zhang et al. 2011, 17298) have come up with a causal model that is thoroughly grounded in empirical data (figure 3.3).

Despite the generic nature of this model, Zhang and colleagues are in a position to neatly illustrate it by Europe's "general crisis" of the seventeenth century. A drop in average temperature around 1560 was imme-

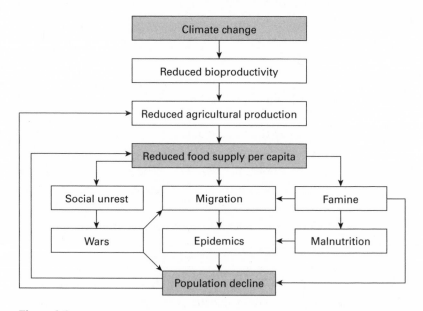

Figure 3.3
Causal pathways from climate change to large-scale human crisis

diately followed by a reduction of bioproductivity, which negatively affected agricultural yields and thus food supply per capita. Over the next thirty years or so, this was followed by cascading escalations of social unrest, migration, famine, war, epidemics, and widespread malnutrition. From 1618, the crisis culminated in the Thirty Years War. Subsequent warfare, together with famines and epidemics, led to a considerable shrinkage of the European population (Zhang et al. 2011).

Data are particularly good for Europe between 1500 and 1800, but the model can also be tested against data from the Northern Hemisphere more generally between 1200 and 1800. Overall, the expectations derived from the model are largely confirmed. As Zhang and colleagues have observed in their earlier piece (2007), it is highly surprising to observe similar macro-patterns for regions as disparate as Europe and China at a time when both areas were largely detached from one another both economically and politically. The authors argue that this synchronicity can hardly be explained unless one assumes social mechanisms triggered by the same kind of climatic stresses.

It is important to note that the model has been developed to account for the social and political effects of a cooler climate in the Northern Hemisphere. Does it also apply to global warming, both in the North and in the South?

The answer is yes. Regardless of the nature of a climatic forcing, the decisive issue is whether or not climate change reduces agricultural production and thereby food supply per capita. If the climatic forcing is strong enough to do that, then the model would predict the same kind of social mechanisms to be triggered. In fact, while Zhang and colleagues have shown that social and political dislocations in the temperate regions of the Northern Hemisphere are mostly associated with climatic cooling, others have demonstrated that the opposite holds for the tropics where warmer El Niño years have always been, and are still, associated with serious social and political trouble (Fagan 2009; Hsiang, Meng, and Cane 2011).[8] From all of this, it seems fair to conclude that global warming of the scale associated with future climate change would have negative effects comparable to those studied by Zhang and colleagues with regard to climatic cooling.

Despite a certain dose of environmental determinism, the model is not linear but shows interaction effects and systemic feedback (symbolized by multiple arrows). What it does not show is the reason for the time lag between the immediate impact of climate change around 1560 and the broader social and political distress in the seventeenth century. Apparently, there were social buffer mechanisms enabling adaptation in the early stages of the crisis. Initially, rulers may have been able to control food riots and contain interstate conflicts from above. From below, rural communities may have been able to fall back on "starvation food." Over time, however, the crisis must have eroded the ability of both rulers and commoners to cope with the situation.

This can be fruitfully compared to the present situation, where industrial society is still able to mitigate the worst effects of climate change, albeit with some cracks already becoming visible. A first serious food crisis occurred in 2008, and another one is currently looming with unprecedented price hikes since 2010. For the time being industrial civilization continues to offer a powerful buffer, but once that buffer becomes brittle there is little to prevent a more serious social and political crisis. In preindustrial times, rulers could repress urban uprisings with a heavy hand, and rural people often silently and invisibly died from hunger. In the present era, however, urbanization and the entitlement mentality resulting from mass consumerism make this kind of social buffer largely unavailable.

Recalling Distant Memories

The causal model presented in figure 3.3 operates on a high level of abstraction. It tells us relatively little about the specific social pathways by which climate change translates into systemic human crises.

To gain a more qualitative understanding of the social and political effects of climate change, let us consider two sets of illustrative case studies. They are selected to address the following question: What are the factors determining the social and political resilience against, or vulnerability to, climate change? That is, what determines whether a society or civilization is able to come up with adaptive social and political responses, or whether it is bound to enter an existential crisis and, eventually, to face large-scale collapse?[9]

My first set of cases is drawn from the ancient Near East, 11,000–1000 BC. During these ten millennia, Mesopotamia and its environs were challenged by four significant climate change episodes. While the responses to the first couple of episodes were largely progressive and led to civilizational upgrades, in the third and fourth episodes most of the Near East's highly advanced civilizations were overwhelmed. My second set of cases is about the medieval Norse settlements on Iceland and Greenland, circa 900–1500 AD. In contrast to the advanced civilizations of the ancient Near East, the focus here is on marginal human habitats. The settlement in Iceland was flexible enough to weather the transition from the Medieval Warm Period to the Little Ice Age, whereas the settlement in Greenland was less adaptable and collapsed.

To my knowledge, these are the best historical cases available to study the social and political effects of climate change. The ancient Near East is a particularly appropriate historical proxy because, since then, there have not been any cases of societies affected by abrupt climate change of a magnitude remotely comparable with what is expected for the twenty-first century. Nevertheless, there is a problem with going back as far as ancient Mesopotamia: while the quality of historical and archeological sources is remarkable for advanced civilizations, it is largely insufficient to trace the effects of abrupt climate change on peripheral communities. To understand the latter, it is necessary to turn to relatively more recent and comparatively weaker episodes of climate change. The history of the Norse settlements on medieval Iceland and Greenland fulfils this purpose.

There are of course numerous other cases where, according to a burgeoning literature, societies and civilizations have entered a crisis and/or faced collapse due to abrupt climate change (Fagan 2004, 2008, 2009; Diamond 2005; Anderson, Maasch, and Sandweiss 2007; Ponting 2007; Chew 2007, 2008; Whyte 2008; Behringer 2010; Mazo 2010, 43–72). This includes the demise of Mayan civilization, the collapse of pre-Columbian pueblo cultures in what is now the southwestern United States, and even the demise of Angkor Wat in Cambodia (deMenocal 2001; Diamond 2009; Buckley et al. 2010; Aimers and Hodell 2011). All of these cases are interesting, but I have selected only the most compelling ones: the ancient Near East and the medieval Far North.

The Ancient Near East, 11,000–1000 BC

One conceivable response to future climate change is the implementation of complex solutions such as various forms of technological innovation and/or a global carbon regime. A possible alternative might be reducing complexity, for example by relying on "community solutions" at the local level and thus striking a new balance between environmental conditions and the complexity of societal organization. If neither of these voluntary options is available, systemic collapse is looming as another possibility—presumably followed by the forced adoption of lower-complexity solutions after some turmoil.[10]

To illustrate this point, it is instructive to look at the first ten millennia of Old World civilizations. The geographical focus is on ancient Mesopotamia, but with an eye to synchronous events in the neighboring regions of the Near East. The time period stretches from the so-called Younger Dryas up to about 1000 BC. The Younger Dryas was a period of very cold and dry climatic conditions, lasting from about 10800 to 9500 BC. According to a recent survey (Roberts et al. 2011), it was followed by a fairly stable and benign climate in the Near East for the next six millennia. The subsequent 3,000 years were punctuated by three somewhat cooler and significantly more arid phases, culminating in multiyear droughts. They occurred toward the end of the fourth, third, and second millennia BC (3300–3000; 2500–1950; and 1200–850).[11] (See table 3.1.)

My assessment is related to Joseph Tainter's theory of the emergence, survival, and collapse of complex societies (1988, 2011). According to his theory, the fate of societies depends on their ability to adapt to emerging challenges either by an upgrade or by a voluntary downgrade of their

Table 3.1
The Ancient Near East, 11,000–1000 BC

Climatic stress	Dominant response	
10,800–9500 BC Younger Dryas	More complexity	Agricultural revolution
3300–3000 BC	More complexity	Rise of urban culture
2500–1950 BC	Temporary increase in complexity, followed by systemic collapse	Northern Mesopotamia: rise and fall of the Akkadian Empire Southern Mesopotamia: rise and fall of the Third Dynasty of Ur
1200–850 BC	Collapse	"Dark ages" all across the Old World

systemic complexity. In general, upgrades are the preferred option. They are particularly rewarding at the early stages, when the marginal cost of higher complexity is low. Later on, the growing marginal cost of complexification makes comparable upgrades gradually more expensive. The strategy becomes entirely punitive at the final stages, when the return on investment in further complexity is negative. Tragically, however, the alternative option of voluntary simplification is hardly available because advanced civilizations are not "downward compatible." They are incapable of a planned reduction of their level of complexity because the existing complexity represents indispensable solutions to real problems. Consequently, involuntary collapse is the only way for the fragments to reach a new equilibrium.

While Tainter's theory was not originally formulated with climate change in mind, the history and archeology of Mesopotamia and neighboring regions suggest that there were significant environmental stresses from climatic changes and the concomitant droughts, forcing the ancient civilizations of the Near East initially to increase their complexity and later to collapse (Weiss 2000; Thompson 2006; Ur 2010). Complex solutions were feasible at the earlier stages of development when existing complexity was still relatively low, while collapse occurred later and took highly disruptive forms such as large-scale imperial breakdown and the abandonment of urban settlements, in some cases followed by "dark ages" lasting for centuries (Chew 2007).

Around 10,000 BC, while Mesopotamia was still only sparsely populated by hunter-gatherers, people in adjacent areas of southwest Asia were already transitioning toward farming (Fagan 2004, 79–96).[12] This "agricultural revolution" is associated with the Younger Dryas (10,800–9500 BC), a period of extremely cold and dry climate reminiscent of earlier ice ages. During that period, hunter-gatherers in southwest Asia responded to the shrinkage of their land base with the introduction of plant cultivation and animal husbandry, which enabled them not only to maintain their population level but even to multiply it by orders of magnitude. This increase in social complexity fits Tainter's theory because it was induced by significant hardship.[13]

Let us move to Mesopotamia (Fagan 2004, 127–137; Ur 2010). For the first two to three thousand years after the eponymous "land between the rivers" had been settled by farming communities, rising population levels gradually imposed an intensification of agriculture around growing human settlements such as Tell Brak in northern Mesopotamia and, most famously, Uruk in southern Mesopotamia. This included increasingly complex farming organization based on artificial irrigation, especially in the south. Rising inputs of communal labor were necessary to counter the effects of land degradation, and most notably the salinization of land and the siltation of irrigation channels.

While the transition to advanced irrigation agriculture was slow and gradual, and probably related to population pressure, the decisive shift from large rural settlements to genuine urban centers was associated with adverse climate change (Brooks 2006, 2010; Kennett and Kennett 2007). Toward the end of the fourth millennium (3300–3000 BC) a period of cooler climate led to serious multiyear droughts on a diminishing agricultural base. This in turn led to a variety of local reactions, such as the abandonment of proto-urban settlements in northern Mesopotamia and a shift of marginal communities to pastoralism, but circumstances particularly rewarded settlements such as Uruk, which by 3100 had transitioned to unprecedented forms of urban complexity.

Urban centers were characterized by high levels of centralization and bureaucratization, with hieratic rulers and elaborate temple bureaucracies overseeing the management of advanced irrigation technology. This made it possible to accommodate rising populations, including an uprooted workforce of economic migrants who had been forced by droughts to abandon their native lands. Rulers were enabled by their control of the granaries and other strategic resources to engage in long-distance trade, while redistributing some of the harvest as standardized food rations to agricultural workers. Thus, climate change in the late fourth millennium was the "midwife" of Sumerian civilization. "In its earliest iterations, the Mesopotamian city was a unique way of responding to environmental crisis" (Fagan 2004, 140).

The next climate crisis occurred around 2500–1950 BC, leading to multiyear droughts from about 2200 BC. An initial response was again social and political centralization, with the Akkadian Empire taking over the entire Mesopotamian system of city-states, including their economy. The organization of agriculture, the distribution of the harvest, and international trade now took place on the basis of imperial bureaucratic control. This time, however, complexification turned out to be unsustainable and the Akkadian Empire collapsed around the year 2150 BC.

The same applies to the Third Dynasty of Ur, which subsequently to the Akkadian collapse acquired a hegemonic position in southern Mesopotamia. The population of southern Mesopotamia swelled due to environmental migrants from the north, including a herding people known as the Amorites. To keep abreast with the problems, Ur's rulers tried a mix of aggressive agricultural intensification, quasi-imperial overstretching, and migration control—even constructing a 180-kilometer wall called the "Repeller of the Amorites." Ultimately, however, their failure tragically led to the demise of their dynasty and the collapse of Ur (Weiss et al. 1993; Weiss 2000, 88–89).

The calamity was associated with the demise of numerous urban centers in the whole of Mesopotamia (Ur 2010). It was transmitted by an explosive mix of economic contraction, trade collapse, regime change, and conflict with migrant invaders from the hinterlands. It is important to note that these invaders, who are sometimes held responsible for the collapse, were actually following a mixture of "push" and "pull" factors. They were displaced from their ancestral land due to increasing aridity, and they were attracted by the unusual military weakness of large urban centers (Thompson 2006).

This was in parallel with synchronous climate-change-induced crises in the neighboring regions (Weiss and Bradley 2001; Lawler 2008). In Egypt, the Old Kingdom disintegrated and collapsed in 2134 BC, with the Middle Kingdom to emerge only after a 170-year intermediate phase. The irreversible collapse of the so-called Harappan civilization in the Indus Valley around 1900 BC may also be related to climate change. In Mesopotamia, the short-lived Akkadian Empire and hegemony of Ur were succeeded by a return to the old system of rival cities, initially at a far lower level of complexity. The Assyrian King List records "17 kings living in tents" for this period, presumably overseeing pastoral societies (as quoted in Weiss 2000, 88).

The next arid phase, from 1200 to 850 BC, coincided with the end of the Bronze Age and with the advent of the Iron Age, often separated by centuries of "dark ages." In Mesopotamia, there was "only" a crisis of various empires in the south and some urban collapse in the north (Ur

2010). In Greece, however, the dark ages following the collapse of the Mycenaean and Cycladic cultures lasted for several centuries. In Egypt, the New Kingdom similarly collapsed around 1070 BC, under the weight of "barbarian" attacks by environmental migrants from marginal agricultural and pastoral areas. The Egyptian Empire did not fully recover from this until 715 BC.

In sum, over a period of ten thousand years, stresses such as population pressure and land degradation placed a constant incentive on the ancient Near East to increase social and political complexity. The major breakthroughs, however, occurred during periods of climate change and related multiyear droughts when adaptation became a matter of civilizational survival. Tragically, adequate complex solutions were not always available and/or sustainable. They were feasible in the first two climate crises (10800–9500 BC, the Younger Dryas; and the period from 3300–3000 BC), whereas the third and fourth climate crises (2500–1950 BC; 1200–850 BC) led to serious cases of socioeconomic and political collapse.

Overall, this supports three general conclusions. First, Joseph Tainter's (1988, 2011) theory of the emergence and collapse of complex societies stands corroborated. At the early stages of development, societies are able to respond to existential challenges such as climate change with civilizational upgrades. As their complexity increases, however, the diminishing returns on investment in further complexity disenable societies from following the same route. This is not to deny that theories of societal collapse remain controversial (Lawler 2010). Nevertheless, expectations from Tainter's theory are largely confirmed by the archeological evidence from the ancient Near East.

Second, when complex societies collapse, they do not disintegrate into atoms but into societal fragments such as tribes, which are likely to regroup after a while. Nevertheless, the recovery of collapsed civilizations is far from easy. In many cases it occurs only after protracted civilizational interludes, or "dark ages." One reason is that the recovery of depleted natural resources takes time. For example, it took time for Mesopotamian agriculture to recover after the demise of the Akkadian Empire and the Third Dynasty in Ur because unsustainable land use had led to irrigation-induced salinity (Montgomery 2007, 39–40). After the collapses of 1200–1100 BC in particular, the "dark ages" in the Eastern Mediterranean lasted for several centuries.

Third, Tainter's theory primarily applies to advanced civilizations that were at the forefront of complexification, as well as their immediate emulators. Other groups remained "primitive" and continued to lead a marginal and precarious existence. In ancient history, our qualitative knowledge about such peripheral communities is limited. Since they have left few cultural artifacts and no written records, their social and political reactions to climate change are difficult to establish. Culturally marginal groups typically show up only as "barbarians at the gates" when apex civilizations have become brittle. This is almost certainly a distorted view.

The Medieval Far North, 900–1500 AD

On a human level, people from marginal communities are as important as those from advanced civilizations. But it is very hard to study them in prehistoric settings, where they left few cultural artifacts and no written records. It is much easier to study them in historical settings, where even marginal communities may be literate and reports from external observers abound. Therefore, let us move a few millennia ahead and focus on a systematic comparison of two highly instructive medieval examples: the Norse settlements on Greenland and Iceland between 900 and 1500 AD. (See table 3.2.)

Iceland and Greenland were first settled by Vikings in the ninth and tenth centuries, respectively. This was during a period of mild climate, known as the Medieval Warm Period (Fagan 2008). From a European perspective both colonies were marginal, with population sizes of around fifty thousand each. Their inhabitants relied mostly on agriculture, plus hunting in the case of Greenland and fishing in the case of Iceland. Both colonies became Christian around 1000 AD and maintained close cultural and trade links with Norway, to the point of formally submitting to Norwegian rule in the thirteenth century.

Despite these similarities, their fate was remarkably different. Iceland remains populated by descendants of the original Norse settlers up to the present day. By contrast, the Norse colony in Greenland collapsed in the fourteenth and fifteenth centuries, after the end of the Medieval Warm Period and at the beginning of what climate historians call the Little Ice Age (B. Fagan 2000).

Table 3.2
The Medieval Far North, 900–1500 AD

Climatic period	Iceland	Greenland
Medieval Warm Period	Norse colonization from 874	Norse colonization from 986
Little Ice Age from ca. 1350	Crisis and adaptation	Crisis and collapse Northern Settlement ca. 1350 Southern Settlement ca. 1450

The Little Ice Age was a period of cooling and increased climatic variability. According to the classical definition by climate historian Hubert Lamb, it was the period when "not only in Europe but in most parts of the world the extent of snow and ice on land and sea seems to have attained a maximum as great as, or in most cases greater than, at any time since the last major ice age" (Lamb 1977, 461–462). In the European case, the Little Ice Age started around 1350 and lasted until about 1850. Overall, there were two episodes of significant cooling of a magnitude of about 1°C in mainland Europe and 2–3°C in the North Atlantic: one at the beginning of the Little Ice Age, in the fourteenth century, and the other from the sixteenth to the nineteenth century. It is important to note that there were important variations of this pattern both in Europe and in other parts of the Northern Hemisphere (Mann et al. 2009).

Let us first consider the colony in Greenland (Diamond 2005, 211–276; see also Whyte 2008, 114–120; Conkling et al. 2011, 130–140), with its two settlements on the West Coast: one to the far south, and one slightly further north.[14] During an early episode of multiannual cooling from 1343 to 1362, the northerly settlement collapsed around 1350 after the farmers had eaten their livestock and even their dogs, and were literally freezing to death. The southerly settlement struggled on for another century, but gave way around 1450 when climatic conditions had become exceedingly hostile to livestock farming and traditional Norse hunting techniques, and when seafaring had become almost impossible due to pack ice. While Greenland remained sparsely populated by Inuit hunters, this was the end of European occupation until Danish re-colonization in the eighteenth and nineteenth centuries.

Norse Greenland was agrarian but not primitive. Its rural economy was tightly integrated with Europe, and its population relatively stratified. The culture of the settlers was staunchly European and, from AD 1000, they were Catholic. They closely attended to the institutions and practices of their Norwegian motherland. While this made them culturally more cohesive, it also made it difficult for them to adapt to environmental change. Had it not been for their ethnocentric mindset, the Greenland Norse could have learned from the Inuit how to survive in a colder environment. To their own misfortune, however, they stuck to Norwegian-type agricultural lifestyles and hunting techniques that became increasingly dysfunctional as the climate got colder.

The agricultural lifestyle, hunting techniques, and food habits of the Greenland Norse were woefully ill adapted to the longer winters and shorter summers of the Little Ice Age. Grazing sheep and cattle became more difficult due to the shortening growing season. Hunting for seals and caribou also became harder. Despite these problems, there is no evidence that the Greenland Norse ever tried to imitate the superior hunting techniques of the Inuit. Incredible as it may sound, they even preferred starvation to eating fish. This self-imposed taboo sounds crazy in an environment teeming with excellent fish such as cod and salmon. And yet no fish bones or other significant evidence of fishery and fish consumption have been found in archeological excavations.

Thus, while climatic and environmental change was the prime mover leading to the demise of Norse Greenland, cultural inflexibility was a crucial intervening factor. Overall, a "combination of climatic deterioration plus environmental degradation . . . and an inability to be flexible and adaptable in the face of environmental stress seems the most likely set of circumstances which finally pushed the colony over the edge" (Whyte 2008, 119).

Next, let us consider the parallel case of the Norse colony on Iceland (Diamond 2005, 197–205; see also Karlsson 2000). Despite severe stresses from climate change during the Little Ice Age, Iceland's population declined but did not collapse. This is remarkable because, all things considered, Iceland was just as vulnerable to climate change as Greenland. This is not to deny that temperatures were slightly milder in Iceland than in Greenland, and Iceland benefitted from more accessible trade routes.

But Iceland was agriculturally even more vulnerable than Greenland and therefore might easily have succumbed to the Little Ice Age. The main challenge was soil degradation due to deforestation and overgrazing. On both islands, the Norwegian-style agrarian way of life pursued by the settlers caused considerable soil erosion. But while in Greenland soil degradation became an existential problem only at the onset of the Little Ice Age, volcanic Iceland continuously suffered from very serious land degradation ever since the very first century of colonization (Whyte 2008, 120–124; Montgomery 2007, 224–228).

Altogether, the Norse colony in Iceland was endangered in much the same way as its counterpart in Greenland. For example, Iceland's chronicle reports a great winter famine in 975–976: "Men ate raven then and foxes, and many abominable things were eaten which ought not to be eaten, and some had the old and helpless killed and thrown off the cliffs" (Jones 1986, 182). This was just one famine in a single winter, centuries before the Little Ice Age led to extremely tough winters and short summers for many consecutive years (Patterson et al. 2010).

So why did the Norse colony in Iceland survive the Little Ice Age, despite the fact that population declined and marginal farms were abandoned? Apparently, the key factor distinguishing Iceland from Greenland was a greater cultural flexibility and willingness to reform environmental management. To prevent further land degradation, farmers "stopped keeping environmentally destructive pigs and goats," and "sought to reach agreement on the maximum number of sheep that each communal pasture could support, and how that number was to be divided among sheep quotas for the individual farmers" (Diamond 2005, 201). Unlike the Greenland Norse, Icelanders did not stubbornly stick to a dysfunctional agrarian lifestyle but reduced their dependency on livestock farming. To make up for the losses, from the fourteenth century they increasingly relied on fishery. Fish became more important not only for domestic consumption, but also as a profitable export staple. Dried cod, or stockfish, replaced wool as the most important good for trade with Europe (Karlsson 2000, 106–110).[15]

In this section, two cases have been considered plus an implicit third case. The Greenland Norse failed to adapt to deteriorating climatic conditions, and ultimately collapsed. The Icelanders were considerably more adaptive and survived, although even they faced severe economic and demographic crises during the so-called Little Ice Age. Only the Inuit in Greenland were thriving; indeed the colder it became, the more they thrived, because their hunting lifestyle was uniquely adapted to freezing Arctic conditions.[16] All of this shows how crucial it is for societies in marginal habitats to be adaptable to changing environmental conditions.

Glimpsing the Future

Despite the fact that climate change is expected to be more severe than any previous climatic shock since the end of the last ice age, the twenty-first century is not the first time humanity, either in its entirety or in part, has been confronted with serious climatic stresses. To gain the necessary analytical leverage to understand the social and political effects of future climate change, we have examined climate change episodes from the ancient Near East and medieval Far North.

The first set of cases, drawn from the ancient Near East, suggests that progressive adaptations to climate change are more likely at the early stages of a civilization when there are large unexploited gains from investment in higher complexity. As the marginal cost of complexity increases, similar upgrades become more difficult and, at some point, impossible. Voluntary simplification is usually not an option, as the current level of complexity is needed to cope with existing problems. Therefore, involuntary collapse is often the only way for the fragments to enter a new equilibrium at a significantly lower level of social and political complexity.

The second set of cases, drawn from the medieval Far North, suggests that the survival of marginal communities depends on their cultural flexibility, and thus their resilience to changing environmental circumstances. While industrial civilization as a whole may be unable to adopt adequate complex solutions to deal with climate change, marginal communities in the so-called developing world may still be at a level of complexity where adaptive solutions are possible. To a significant extent, the ability of such communities to adapt to climate change in the face of deteriorating living conditions will depend on their flexibility.

For advanced industrial countries, the decisive question is whether industrial civilization is at a level of complexity where it is still possible to come up with progressive solutions to cope with stresses such as climate change. How easily can industrial society develop adequate technological responses to climate change and agree on viable multilevel governance schemes to reduce carbon emissions? Can it mobilize complex solutions to prevent its own collapse, or is there an "ingenuity gap" between the magnitude of the challenges ahead and its limited ability to find adequate solutions (Homer-Dixon 1995, 2001, 2006)?

Today, technical ingenuity requires escalating amounts of money and time. There is evidence to suggest that research and development is growing increasingly expensive, and is taking more and more time (Strumsky, Lobo, and Tainter 2010).[17] In the crucial energy sector, technological innovations are almost prohibitively expensive and take considerable time to develop and roll out (Hirsch, Bezdek, and Wendling 2005, 2010).[18] While the prospects for technical ingenuity are limited, the chances for social ingenuity are not much better. The increasing international disunity since the 2008 financial crisis and the 2009 collapse of the Copenhagen summit on climate change suggest that globally adequate multilevel governance solutions are politically unavailable.

If technical ingenuity and political complexification do not hold the key to a solution for our global problems, then voluntary simplification may offer an alternative. This is the hope of those promoting "transition towns" and other local "community solutions" (Hopkins 2008; Murphy 2008; De Young and Princen 2012). Their strategy is for environmentally conscious individuals to get together and work out sustainable low-tech lifestyles in order to reach a new balance between environmental conditions and social organization. Indeed, a case can be made that in the long run this may be the only viable solution (Greer 2009).

Alas, community solutions cannot work without considerable solidarity and social cohesion. This is precisely what is lacking in rich industrial countries, where social capital has been undermined by the effects of economic affluence and mass consumerism (Putnam 2000). Under such circumstances, the deliberate investment in community solutions can only be a fringe phenomenon. A genuine communal revival is not likely to happen unless and until it is forced by systemic collapse (Greer 2008; Kuecker and Hall 2011). With due respect to environmentally conscious individuals, it may take a dreadful period of "dark ages" to force the fragments of industrial society to find a new sustainable equilibrium (Chew 2007, 2008). Moreover, with due respect to well-meaning communitarians and local activists, modern civic achievements such as multiculturalism and gender equality may be lost when industrial civilization is replaced by land-based neo-traditionalist lifestyles (Quilley 2011, 2013).

A related issue is whether it will ever be easy, or even possible, to reboot complex industrial civilization after a deep global crisis—whether forced by climate change or by something else. If we assume a large-scale collapse, then a recovery will certainly be difficult. For example, it would be hard to resume advanced mining operations after a serious deterioration of the industrial base. It would also be difficult to reactivate networked infrastructure, including advanced information technology, once their industrial underpinnings are gone.[19]

For people in poor developing countries, there is good news and there is bad news. The bad news is that poor countries are highly vulnerable to climate change due to intense population pressure and a limited ability to mobilize complex industrial solutions (Homer-Dixon 1995). If we assume a breakdown of international humanitarian aid, enormous suffering will be caused by the fact that population levels in many poor countries are way beyond what these countries could sustain on their own. From this perspective, in poor countries climate change is likely to lead to enormous human suffering and violence.

Even today, the risk of violent civil conflict in tropical countries is twice as high in warm and dry El Niño years as in cooler and wetter La Niña years (Hsiang, Meng, and Cane 2011). Especially in Africa, higher temperatures and greater rainfall variability are associated with more frequent civil wars and small-scale conflicts (Burke et al. 2009; Hendrix and Salehyan 2012; Raleigh and Kniveton 2012). While some of these claims about the climate/conflict nexus remain debated (e.g., Buhaug 2010), plain commonsense would seem to suggest that people in poor countries are particularly vulnerable to climate change.[20]

In the long run, however, there is also significant good news for people living in poor countries: they should be able to preserve more of their way of life than people living in rich industrial societies. If we imagine a systemic demise of world industrial civilization, people in poor countries may often be in a better position to recover from mayhem than individuals living in rich countries. The reason is that the solidarity and social cohesion of many communities in poor countries is tighter, which makes them more resilient compared to individuals from wealthy consumer societies. Despite all the hardship resulting from the negative impacts of climate change and the breakdown of humanitarian aid, this community resilience is likely to make it easier for many groups of people in poor countries to brave a systemic crisis of the world system (Kuecker and Hall 2011).

Even in poor countries, the level of community resilience, and by extension the expected level of post-crisis adaptability, is bound to vary from group to group. Thus, cultural adaptability clearly differs for farmers and herders; Hindus and Muslims; and urban and rural populations. Moreover, different groups are likely to recur to different adaptive mechanisms. Some groups may peacefully and silently adapt; others may migrate; and yet others may rather start fighting (WBGU 2008).

Other things being equal, the consequences of climate change in poor countries are likely to be more negative in hotspots where population density and other environmental vulnerability indicators are high (K. Smith 2011). They may be more benign where there are high levels of household and community resilience, as well as a good quality of governance and low levels of communal violence (Busby et al. 2012). People in rich industrial countries are likely to have a greater buffer at the initial stages of the crisis, but may have a harder time adapting later on.

Despite these important differences between rich and poor countries, modern industrial civilization constitutes a uniquely closed social and ecological system (the "Spaceship Earth" worldview formulated by Kenneth Boulding and popularized by Buckminster Fuller in the 1960s). This is different from most of human history when civilizations were based on agriculture and either surrounded by competing civilizations or by barbarian hordes living in the "wilderness." When facing serious problems, ancient civilizations were sometimes able to become more sophisticated. Another strategy available to them was to increase their resource base by occupying and exploiting new land at their periphery.[21]

When everything else failed, one result was a great migration. People from marginal communities would exploit the relative weakness of the apex civilization, while people from the apex civilization would leave their areas of origin in search of better habitats. Similar kinds of migration are likely to remain a typical response to climatic stresses (Piguet, Pécoud, and De Guchteneire 2011). But whereas the survivors from the collapse of ancient civilizations, such as the Mayans in the ninth century, could effectively disperse into the wilderness, habitat tracking is not going to be an adaptive systemic response in a world crowded by seven to nine billion people (Weiss and Bradley 2001, 610).

Bottom Line

Industrial civilization, in all its complexity and including the associated cosmopolitan values of tolerance and cultural diversity, has been a source of strength during conditions of growth as they have prevailed for the last two centuries. In a period of stress, whether due to climate change or any other massive constraint, this may turn into a liability because consumerist society suffers from excessive complexity and a fatal lack of social cohesion.

4

When Energy Runs Short

There is no denying that industrial society runs on energy, and especially on oil, so a serious fuel shortage can bring it to the brink. Most readers will associate fuel shortages with the oil crises of the 1970s, but it is important to note that in neither of these cases did the supply shortfall last for more than a few months or amount to more than 7 percent of global oil consumption. Now imagine a piecemeal but steady reduction of world oil supply by, say, 3 or 4 percent per year for a couple of decades. Surely this would have momentous social and political effects.

Today, serious energy scarcity can result from two different contingencies. The first contingency is *climate-change-induced energy scarcity*. Climate change poses a constraint on how much CO_2 can be emitted, so rationing fossil fuels may become necessary to slam the brake on runaway climate change. The second contingency is *energy scarcity caused by resource depletion*. Most notoriously, a peak and subsequent decline of world oil production may lead to a serious shortage of the key energy resource fueling industrial society.

Of the two contingencies, my focus is on serious energy scarcity in the wake of peak oil. But I surmise that climate-change-induced energy scarcity would have similarly disruptive effects, unless we assume that industrial societies are willing and able to voluntarily reduce their fuel consumption in a gradual and carefully planned manner—which, for now, is clearly not happening (see box 4.1).

In this chapter, I ask a "what if" question: What is likely to happen in different parts of the world if peak oil leads to massive energy scarcities? As a baseline, I hypothesize a decline of global oil production by 2–5 percent per year for a couple of decades, after a few years on a bumpy plateau. In line with most of the peak oil literature, I further

Box 4.1
Climate-Change-Induced Energy Scarcity

> The current stalemate in climate policy suggests a conceivable scenario for climate-change-induced energy scarcity. A major natural calamity might wake up decision makers to the reality of runaway climate change. For example, sudden desertification in large swaths of the United States and China might induce leaders in these countries to aggressively curtail fossil fuel consumption and, potentially, international trade in high-carbon fossil fuels, notably coal. Together with Western Europe, the United States and China would then constitute a credible core for a global carbon compact. Although the scenario is unlikely and there are no historical precedents for it, and despite the fact that there would surely be countries refusing to join the compact, at some point there may be a manifest choice between frying the planet or going cold turkey on fossil fuels. Frying the planet seems more likely, but going cold turkey is at least a remote possibility that will be explored further below (see box 4.2 on p. 102).

hypothesize that no adequate alternate resource and technology will be available to replace oil as the energy backbone of industrial society.

An event comparable to peak oil has never happened at the global level. The likely effects of peak oil are far above the formal threshold stipulated by the IEA for an international oil supply disruption (7 percent) and also higher than the shortfalls of global oil production during the oil crises of the 1970s (less than 7 percent). Because this is so, I resort to cases where disruptive energy scarcity in the order of 20 percent or more has occurred at the national level. I will argue that studying national cases of disruptive energy scarcity as "proxies" is the best strategy available to gain analytical clarity about the likely effects of peak oil. My analytical focus is on oil importing countries, which constitute the vast majority of states.

My first case is Japanese *predatory militarism* before and during the Pacific War. The fear of resource shortages had played an important role in shaping Japan's imperialist strategy ever since the end of World War I. When a US oil embargo became imminent, in 1941, Japan preemptively attacked the US naval base at Pearl Harbor and radicalized its war of conquest in order to gain access to the rich oil supplies of the East Indies.

My second case is *totalitarian retrenchment* in North Korea after the end of the Cold War. When subsidized deliveries of oil and other vital resources from the Soviet Union were disrupted, the "Hermit Kingdom" reacted in a shockingly reckless way. Elite privileges were preserved in the face of hundreds of thousands of North Koreans dying from hunger. While this may be morally repugnant, it represents another possible reaction to disruptive energy scarcity.

My third case is *socioeconomic adaptation* in Cuba, which was challenged by a similar disruption of subsidized deliveries from the Soviet Union. While this plunged Cuba into a deep crisis, there was no mass starvation comparable to that in North Korea. Instead, Cubans relied on social networks and non-industrial modes of production to cope with energy scarcity and the concomitant food shortage. They were actively encouraged to do so by the regime in Havana.

So-called techno-optimists typically object to "Malthusians" that a global decline of oil production would not only lead to higher prices but also trigger a transition from oil to other energy sources, such as renewable energy or a new generation of nuclear reactors. Appealing as it may be, unfortunately this argument is countered by another historical case study: the South of the United States, or "Dixieland," after the American Civil War (1861–1865).

After the American Civil War, Dixieland was deprived of slaves as the backbone resource of the "Southern" way of life. Without prior knowledge, one would probably expect this to be a clear-cut case of a smooth transition. After all, Southerners only had to look to the northern part of the United States for investment and innovative technologies. Nevertheless, the modernization of Dixieland took a century if not more. Insofar as similar upgrades do not seem to be available in the event of peak oil, there is no reason to be particularly optimistic about a smooth transition to a post-oil (or even post-carbon) society.

So here is my plan for the chapter: After providing a justification for my case-study-based method, I present the historical cases of Japan, North Korea, and Cuba. Each case outlines a specific response to an acute or (in the Japanese case) anticipated disruptive energy scarcity. Next, I formulate some generic hypotheses about the factors that would determine how different parts of the world might react to disruptive energy scarcity. To counter the view that the transition to a post-oil society will be easy, I present my case study on Dixieland. All of this enables me to develop a specific scenario of what might happen in different parts of the world during the first couple of decades of mounting energy scarcity.

How to Study Disruptive Energy Scarcity

My strategy to study disruptive energy scarcity involves deriving analytical insight from historical cases where it has actually occurred, namely when a country or society faced a rapid decline of access to its backbone energy resource. A *backbone energy resource* is an essential resource that more than any other physical resource enables a social, political, and economic way of life. A resource is essential insofar as it is vital for human subsistence and/or economic production and cannot be readily substituted (Ehrlich et al. 1999). Today, our backbone energy resource is oil. A century ago, it was coal. Even further back, it used to be labor.

A *disruptive energy scarcity* is defined as a significant and rapid contraction in the supply of a society's backbone energy resource in the order of 20 percent or more. The 20 percent cutoff is reasonable for my analytical purposes because I am primarily interested in what might happen during the first two decades after peak oil leads to a terminal decline in world oil production. In line with the expectations found in the peak oil literature, I assume a decline of global oil production in the order of 2–5 percent per year (Hirsch 2008; Sorrell et al. 2010a, 4998–4999), presumably after a few years on a bumpy plateau. Despite some geographical variation, an annual reduction of global oil production in the 2–5 percent band would translate into a supply reduction of more than 20 percent within less than two decades.

My focus is on oil importing countries because most of today's advanced industrial societies, with a few notable exceptions, are in that category. Insofar as oil is the energy backbone of industrial civilization, all of these societies are highly vulnerable to a decline of access to oil. In case of a massive oil supply disruption, they are bound to experience a formidable energy crunch.

A critic might object that peak oil is a global event, so historical precedents would have to be found at the global level. The problem with this criticism is that, quite simply, no disruptive energy scarcity has ever occurred at the global level. This is not to deny that the oil crises of the 1970s and other more recent historical oil shocks led to soaring oil prices and had serious economic consequences (Hamilton 2009, 2011, 2013). And yet they do not qualify as disruptive energy scarcities according to the definition given above. They do not even qualify as major energy supply disruptions under the definition used by the International Energy Agency (shortfall of oil supply of 7 percent or more). According to the IEA (2012d) the largest disruption so far was the 1979 Iranian revolution with a 5.6 percent shortfall, followed by the 1990–1991 Gulf War and the oil crisis of 1973–1974 with a 4.3 percent shortfall each (see also Maugeri 2006, 103–119).

In the absence of historical precedents at the global level, we are forced to look to specific countries or societies for suitable proxies. Even at the domestic level, however, we find only few precedents. Events like the California electricity crisis of 2000–2001 or the Ukrainian gas crisis of 2008–2009 are minuscule both in scope and duration when compared to the far more restrictive definition of a disruptive energy scarcity given above.

A perusal of the historical record indicates that, until modern times, disruptive energy scarcities were extremely rare. The reason is obvious. In preindustrial times, few societies were either willing or able to run the risk of depending on a backbone resource imported from abroad.[1] Only after the industrial revolution were ever more countries able to dispense with autarky. They could do this because rapidly expanding resource extraction in combination with international trade made external resource dependency compatible with the preservation and enhancement of security, power, and plenty.

As long as industrial capitalism can draw on an abundant resource base, disruptive energy scarcities will occur in two constellations only: situations of siege, and industrial countries that are cut off from foreign supplies. Situations of siege are interesting (see for example Offer 1989, 2000), but I leave them aside here because there is little to indicate that peak oil will result from such a situation. Instead, the best available proxy is industrial countries that suffer a disruptive energy scarcity because they are cut off from foreign supplies.

My sample (table 4.1) comprises all relevant historical country cases of disruptive energy scarcity, as defined above, that can be taken as reasonable proxies for a global situation of mounting energy scarcity after peak oil.[2]

Table 4.1
Case Sample

Case		Challenge
Japan	1918–45	Fear of economic strangulation and fuel starvation
North Korea	1990s ⎫	Massive loss of access to subsidized oil deliveries
Cuba	1990s ⎭	

Admittedly, the cases listed in the table are somewhat remote from the experience of Western liberal capitalist societies. Moreover, while definitely being abrupt, energy shortages after peak oil would make themselves felt somewhat more gradually than in the cases of either North Korea or Cuba.

In a social scientist's ideal world, there should be more appropriate historical cases to study the social and political effects of disruptive energy scarcity. In the real world, however, these are simply the closest analogs for peak oil that have ever occurred. They are not perfect, but I suggest that it is better to cautiously harness the historical precedents available to study the social and political consequences of disruptive energy scarcity, rather than to not study them at all.

Studying historical precedents at the national level is also justified by the fact that energy scarcity can lead to a process of de-globalization, with the world again becoming geographically more fragmented and, thus, less "global." Insofar as globalization has been fueled by increasing inputs of cheap and abundant energy traded as a commodity in a free market, a peak of oil production may undermine the very foundations of the worldwide social, economic, and political normalization processes that have been observed over the past few hundred years. While initially a global peak of oil production would per definition be a planetary event, reactions would increasingly vary in different parts of the world.

An insistent critic might further object that formal modeling is a better strategy to study the social and political effects of disruptive energy scarcity. I respectfully disagree. While formal modeling is no doubt highly pertinent to examine general system dynamics and explore future scenarios, it is an illusion to assume that it leads to more directly applicable and unequivocal results.

Ever since the 1970s, scholars have explored world models to understand the global dynamics of resource depletion and scarcity (Meadows et al. 1972; Council on Environmental Quality and US Department of State 1980; Meadows, Randers, and Meadows 2004). Because the world is composed of diverse socioeconomic and sociopolitical systems, others have constructed models operating at the regional and national level (Choucri, Laird, and Meadows 1972; Mesarović and Pestel 1974; Turner, Keen, and Poldy 2011; Lutz et al. 2012).

All of this is valid, but for two reasons it cannot work for our present purposes. One is that formal models tell us little about the social and political consequences of disruptive energy scarcity. Most of them focus on physical flows and/or economic processes, but the passage from there to the sociopolitical sphere is difficult. Another problem is that the assumptions underlying the models are starkly contested. Some models are based on general equilibrium theory (IMF 2011; Waisman et al. 2012), while others rest on ecological economics (Ayres and Warr 2009). Some take a linear view of technological progress (IEA 2010b, 2012b), while critics emphasize the non-linearities and tipping points (Korowicz 2010). Some models concentrate on material flows (Meadows, Randers, and Meadows 2004), while others take the financial economy into account (Turner, Keen, and Poldy 2011). The relatively simple yet powerful model developed by Jaromir Benes and colleagues (2012) is probably as good as it gets, but still leaves many crucial questions open as the authors readily acknowledge.

The result is that virtually any vision of the future is supported by some formal model. To avoid such indeterminacy, my analysis rests on a historically and empirically more grounded strategy. As mentioned, I examine specific cases when socioeconomic and political systems have actually experienced disruptive energy scarcity. Unlike formal models, such historical precedents can serve as suitable proxies to study the social and political effects of disruptive energy scarcity.

In sum, careful inference from historical cases is the worst research strategy except for all others. Bearing in mind that lessons from the past cannot be directly applied to the future, my strategy is to harness relevant historical precedents. This enables me to extract the variables that are likely to determine how different parts of the world would be affected by disruptive energy scarcity after peak oil. Together with plain historical and political common sense, this in turn enables me to develop a scenario of how they would be likely to respond to the crisis; and what the consequences of both the crisis and the responses might be.

Predatory Militarism: Japan, 1918–1945

In September 1945, defeated Japan was so fuel-starved that it was difficult to find an ambulance with sufficient fuel to transport Premier Tojo to a hospital after his attempted suicide. Pine roots had been dug out from mountainsides all over the country in a desperate attempt to find a resinous substitute to fossil fuel. Much of the Japanese air force and navy had been sacrificed in kamikaze raids, at least in part because there was not sufficient petrol to refuel planes and ships to return from their sorties and keep fighting (Yergin 1991, 362–367).

Ultimately, Japanese fuel starvation was the result of a self-fulfilling prophecy. Originally, it had been the fear of military and economic strangulation by an oil embargo that led Japan to radicalize its strategy of imperial expansion and to fatefully engage in full-blown predatory militarism. This in turn led precisely to the fuel starvation that Japanese planners were dreading so much.

Ever since the late nineteenth century, a modernizing Japan had been committed to military conquest in order to compete against overextended European empires (Beasley 1987; Barnhart 1995). The strategy was to emulate countries like Britain and France in their effort to achieve prosperity, power, and glory by the acquisition of overseas territories. In the absence of adequate power projection capabilities, Japan's strategy was concentrated on East Asia where it won a war against China, conquering Korea and Taiwan (1895), and another war against Russia consolidating its territorial claim on Korea and expanding its influence into Manchuria (1905). Japan's participation in World War I further expanded its overseas territories to encompass former German possessions.

The main lesson the Japanese military took home from World War I was that a country cut off from access to raw materials was bound to lose in a military contest due to a trade embargo. In their view, Germany had lost the war because it did not muster the necessary industrial base or access to foreign markets to achieve wartime autarky. To be prepared for a similar war, resource-poor Japan would have to control access to strategic resources. Only a self-sufficient economic bloc in East Asia would sufficiently prop up Japanese industrial capacity to secure the desired status of a great power. From this perspective, the US-sponsored Open Door policy of free trade in the Pacific was not in Japan's interest (Barnhart 1987, 9–21; Beasley 1987).

While this was the dominant mindset in Japanese military circles, the 1920s saw a number of civilian governments experimenting with liberal internationalism. These experiments came to a halt after the world economic crisis of 1929, however, and its consequences convinced Japanese elites that free trade was not a viable strategy for power and prosperity (Smethurst 2007). To survive as a great power, Japan had to be in a position to control raw materials and markets in its geopolitical region. From the early 1930s, Japan was thus committed to an ambitious strategy of imperial brinkmanship.

To prevent fuel starvation and mitigate external dependency on other strategic resources, Japan started to embark on aggressive military campaigns. After the liberal interlude of the 1920s, the next decade saw the invasion of Manchuria (1931), followed by the invasion of China (1937). The paramount goal was to achieve self-sufficiency in an economic bloc that was later, in 1940, to be proclaimed as the "Greater East Asia Co-prosperity Sphere."

This military escalation led to a painful dilemma. On the one hand, it spurred giant efforts to convert Japan's industrial base into a self-sufficient war economy. On the other hand, it undermined Japan's ongoing attempt to achieve wartime autonomy because, even from the cynical viewpoint of Japanese military planners, targets had not been selected wisely. While Manchuria and the other occupied territories yielded significant quantities of food, coal, and iron, very little oil came from these areas. Thus, instead of becoming more self-sufficient, Japan grew even more dependent on the importation of critical commodities—especially from the United States.

The situation was particularly dramatic for oil, which was indispensable as a military transportation fuel. While Japan could stockpile considerable amounts of petroleum and other strategic resources, such stockpiles would not be sufficient in the event of a protracted war to be fought without foreign oil imports. Since the United States was the dominant producer of petroleum at the time, Japan was heavily dependent on American deliveries. Japan was importing 90 percent of its petroleum consumption, of which 75–80 percent was shipped in from California. For the critically important gasoline, the dependence was even higher (Miller 2007, 156–167; see also Worth 1995, 125–135).

With that in mind, it is easy to understand (certainly not to condone) what happened when Tokyo felt threatened by the specter of a US trade embargo: the limited Japanese onslaught in East Asia degenerated into what was to become the relentless Pacific War. The only alternative to importing oil from the United States was looting it from Borneo and Sumatra in the East Indies. To reduce Japanese vulnerability to a US embargo, a southward advance was thus irresistibly appealing—especially to elements in the Japanese navy.

The idea of a southward advance became ever more compelling after the start of World War II, when increasing demand for raw materials in the European theater led to rising commodity prices on the world market. At the same time the United States, which had hitherto limited its acts to token gestures, gradually began introducing real economic sanctions against Japan in the late 1930s. Given the worsening fuel scarcity and in anticipation of a full-blown trade embargo, the Japanese army began its southward advance. Japan started an offensive in southern China in 1939, and occupied the northern part of French Indochina in September 1940 (Barnhart 1987, 136–175).

After endless vacillations, the full-blown US trade embargo finally came in July 1941. While earlier restrictions on exports could be circumvented, this time a freeze of Japan's financial assets made sure that Tokyo would be unable to purchase oil or any other goods from the United States (Miller 2007).

While the embargo was intended to dissuade Japan from escalating its military operations, it had the opposite effect because Japanese imperialism was already far too entrenched for a climb-down. Tokyo took the embargo as the ultimate confirmation that there was no other choice but to move further southward and to tap the rich mineral resources available in the Dutch East Indies, and particularly the petroleum that was being extracted in the British part of Borneo (Ike 1967; Sagan 1988; Yergin 1991, 305–327).

To secure its flank in the imminent military offensive, the Japanese Navy famously undertook a pre-emptive attack on the US Pacific Fleet stationed at Pearl Harbor. The intention was to roll over East Asia and create an entrenched geopolitical bloc while America was directing most of its attention toward the European theater, and later to negotiate some settlement with the United States from a position of relative strength (Kershaw 2007; Record 2009).

Japan was eventually defeated, but only after a cataclysmic war. From 1945, Tokyo embraced the opportunity to participate in a multilateral free trade regime, although liberal free trade was blended with a strong continuity of state intervention. But that is a different story. What matters for the present analytical purposes is that, during the 1930s, resource-starved Japan was prompted by the specter of fuel starvation to scrap the Open Door policy and to try to build a regional geo-economic bloc to prevent strangulation.

Up until the Pacific War, Japan had acted like a belated colonial power struggling to carve out its own "place in the sun." It was only the fear of economic strangulation and fuel starvation concomitant with the US trade embargo that triggered the worst of Japan's predatory militarism.

Totalitarian Retrenchment: North Korea, 1990s

Whereas Japan in the 1930s and early 1940s went on conquest to assert its status as a great power and to secure access to vital supplies, during the 1990s the totalitarian regime of North Korea, formally known as the Democratic People's Republic of Korea, retrenched in order to preserve elite privileges. The main problem was a severe disruption of subsidized oil deliveries after the demise of the Soviet Union. As a consequence, the Great Famine of 1995–1998 led to the starvation of six hundred thousand to one million people, or 3–5 percent of the North Korean population (Goodkind and West 2001, 234).

From 1948 to 1994, North Korea's "Great Leader" Kim Il-sung presided over a country with a relentless Stalinist regime, a Soviet-style industry, and a toxic agriculture. The regime was more totalitarian and reclusive than anywhere else in the Eastern Bloc. In 1990, estimated per capita energy use was twice as large in North Korea as in China and more than half that of Japan. Life expectancy was high, and over 60 percent of the population was urban (Eberstadt 2009, 127–157; Williams, Hippel, and Nautilus Team 2002, 112; see also Park 2002).

The industry, which was based on coal and steel, was as wasteful as in any other Soviet country. In line with the national ideology of self-reliance (*juche*), up until the 1980s the regime had heavily invested in coal mines and hydropower to satisfy the country's enormous energy needs. Furthermore, Pyongyang had developed a toxic "modern" industrial agriculture to feed the highly urbanized North Korean population. Farming was based on irrigation, mechanization, electrification, and the prodigal use of chemicals.

All of this came to naught with the demise of the Soviet Union when it turned out that access to oil was the Achilles heel of the North Korean economy. Since North Korea does not possess any proven reserves of petroleum, oil and other commodities were mostly imported from the Soviet Union and China for "friendship prices," in exchange for political allegiance. The North Korean economy was geared toward domestic consumption and did not produce any competitive export staple, except for some advanced weapon systems. The country was thus running a permanent trade deficit, and its economy was not in a position to generate the revenues necessary to substitute for the subsidized energy inputs that were being delivered by external protectors.

The situation came to a head in 1991, when post–Soviet Russia stopped subsidized exports of oil and other vital goods to North Korea. Two years later, Russian exports to North Korea were down by 90 percent (Haggard and Noland 2007, 27–32).[3] This had dramatic effects. While the North Korean regime reserved most remaining fuel for the military, the rest of the industry nearly collapsed and agricultural production languished around subsistence level. Almost immediately, Pyongyang was forced to launch a "Let's Eat Two Meals a Day" campaign. In 1994, when Kim Il-sung bequeathed leadership to his son Kim Jong-il, a serious food crisis was looming. After a series of decent harvests due to favorable weather conditions in the early 1990s, severe floods and droughts led to the Great Famine between 1995 and 1998 (Haggard and Noland 2007, 73–76; Schwekendiek 2011, 55–60).

The North Korean Great Famine is a paradigm example of how the shortage of a backbone energy resource such as oil can have momentous systemic ripple effects. To begin with, agricultural machinery depended on oil. Without fuel, tractors and other machines were not running. The next problem was transportation. Fuel was needed to transport fertilizer and other inputs to farms, and agricultural products to urban consumers. Fuel was also needed to ship coal from mines to fertilizer plants, where coal was converted into soil nutrients.[4] Fuel was further needed to get coal to power stations for electricity generation. As a consequence, electricity was yet another problem. Without sufficient electricity, irrigation pumping and electrical railways became intermittent. The intermittency of electrical railways further affected transportation. Without reliable trains, it became even more difficult to bring coal to fertilizer plants or power stations, to transport fertilizer to farms, and to get agricultural products to urban consumers (Williams, Hippel, and Nautilus Team 2002).

Thus, interlocking energy shortages, combined with shortages of industrial inputs and a general decline in infrastructure, produced a dramatic decline in production, and thus an almost hopeless situation. While the entire economy was damaged, the consequences were most dramatic in agriculture and resulted in plummeting food production, considerable loss of arable land, and a rapid depletion of soil fertility. Restoring soil fertility would have required large amounts of lime, which could not be transported without fuel. The regime sent more urban workers and school children to the fields, but this did not compensate for the losses. In a desperate attempt to replace agricultural machinery, most animals for meat consumption were culled and draft oxen slowly became more numerous. But, unlike tractors, working animals compete with humans for food. The energy crisis also compelled many poor people to rely on biomass for cooking and heating. Unlike fossil fuel, however, the extraction of biomass reduces soil fertility, which in turn aggravated the agricultural crisis.

As a result of these and other interlocking vicious circles, the production of rice and maize fell by almost 50 percent between 1991 and 1998 (FAO/WFP 1999). The public food distribution system crumbled. Since distributed food rations were the most important form of payment to workers, this led to a further decline of industrial activity. North Korea was thus compelled to apply for international food aid. After a considerable time lag, the worst starvation was stopped in the late 1990s. But since North Korea's industrial agriculture cannot be restored without a viable energy regime, even today there is still a protracted food crisis with an ever-present risk of further starvation. Because the problem is structural, international food deliveries cannot solve it.[5]

Some scholars have interpreted the Great Famine as a malfunction of North Korea's Stalinist regime (Natsios 2001).[6] Politically speaking, however, this misses the point. Pyongyang's performance is dysfunctional only when measured against Western humanitarian standards. On its own (more cynical) terms, the regime has been incredibly successful in the face of considerable duress. The crisis prompted North Korean elites to abandon the Stalinist path of wasteful industrialism and to administer systemic scarcity instead. In the early 1990s, Pyongyang was facing the same tough choice as any other communist country: regime change, limited reform, or totalitarian retrenchment.

The negative policy choice of totalitarian retrenchment made it possible for the regime not only to survive, but also to avoid an economic and political opening, thus preserving cherished elite privileges. While this has caused tremendous hardship for the masses, elite privileges were effectively secured. While most other communist regimes have either disappeared from the map, like the Soviet Union, or become politically and economically more open, like China, the Democratic People's Republic of Korea has remained steadfast.[7] North Korea has even become a nuclear power, which sometimes enables Pyongyang to extort international concessions. While such brinkmanship may be morally repugnant, Korean-style totalitarian retrenchment represents without any doubt a possible response to a disruptive energy scarcity.

Socioeconomic Adaptation: Cuba, 1990s

Cuba and North Korea have much in common. Both are socialist developing countries and centrally planned economies. During the Cold War, both countries thrived on subsidized imports of oil and other goods from the communist world and had an industrialized agriculture based on the wasteful use of fuel, fertilizer, pesticides, electricity, and other inputs. With the demise of the Soviet Union in the early 1990s, both countries plunged into a deep economic crisis due to the interruption of such subsidized imports.

Until 1989, Cuba enjoyed excellent terms of trade with the Soviet Union and Eastern Europe. Sugar and other export staples were sold for solidarity prices propping up the Cuban economy for political reasons, while raw materials and industrial products were bought for friendship prices equally benefiting the Cuban economy. Such was the largesse of the Soviet Union and its European clients that Cuba gained significant amounts of hard currency from re-exporting oil to third countries. The Cuban industry could also rely on support from the communist world for machines and know-how. The food sector counted on subsidized imports of wheat, milk powder, animal feed, fertilizer, pesticides, and so on.

All of this came to naught in 1990 when preferential trade with the Soviet Bloc collapsed, forcing Cuban leader Fidel Castro to proclaim a national emergency called the "Special Period in Time of Peace" (Mesa-Lago 1993). As a result, Cuba faced an energy supply disruption similar to the one experienced by North Korea. When taking into account the fact that heavily subsidized oil deliveries from China to North Korea lasted until 1993, the Cuban supply shock was even more abrupt and dramatic. Subsidized energy supplies from the Soviet Bloc ceased to 100 percent from one year to the next. The CIA (1996, 9) calculated the decline of Cuban fuel imports between 1989 and 1993 at a whopping 71 percent.[8]

The crisis entirely devastated the Cuban economy. Machines lay idle in the absence of fuel and spare parts. Public and private transportation were in shambles, with people walking and cycling long distances or riding on modified vans called "camel buses." Workers had difficulty getting to their jobs. Factories and households all over the island were struck by rampant and unpredictable electrical power outages (Pérez-López 1995, 138–140).

As in North Korea, the most painful effects were felt in the food sector. From a daily chore under real communism, the procurement of food became a real source of anxiety to consumers. The nutritional intake of the average Cuban, especially protein and fat, fell considerably below the level of basic human needs (Alvarez 2004, 154–169). Consumers resorted to chopped-up grapefruit peel as a surrogate for beef, and some people started breeding chickens in their flats or raising livestock on their balconies (Pérez-López 1995, 138).

Despite such considerable hardship, Cuba was far more resilient than "self-reliant" North Korea. Common people in Cuba were not dying from malnutrition and starvation. Homeless people and gangs of street children, turned into scavengers, were not characteristic features of Cuban townscapes. Nor were violence, crime, desperation, and hopelessness characteristic features of Cuban neighborhood life (Taylor 2009, 144–145).

The Cuban scenario is in remarkable contrast to the situation in North Korea. Although reliable accounts are in short supply, reports from North Korean exiles indicate that during the 1990s everyday life in the so-called Hermit Kingdom was solitary, poor, nasty, brutish, and short (Natsios 2001). As mentioned, famine killed as many as 3–5 percent of the North Korean population. While life was certainly hard during the Special Period, nothing of that sort happened in Cuba.[9]

The immediate reaction of the Cuban regime was predictable: mobilize the masses for food production, and revitalize the state sector. Townsfolk were sent to the countryside for farm labor, but after more than forty years of real communism there was little revolutionary fervor left in the population. Also, the state sector was too sclerotic to be converted from sugar and coffee to potatoes and beans. Despite world market prices for sugar below production costs, state farms continued to produce sugarcane (Burchardt 2000).

The next response of the Cuban regime was cautious liberalization and reform. To begin with, the regime moved from toleration to the controlled legalization of certain black-market and informal-sector activities. To attract hard currency, the country was cautiously opened to Western tourists. The US dollar was legalized as a parallel currency. Control over numerous state farms was partly devolved to the employees and management. All of this contributed to a burgeoning informal and semi-informal sector, which quickly took on its own dynamic and significantly contributed to the provisioning of the Cuban population (Pérez-López 1995; Padilla Dieste 2002).[10]

This strategy was not only more flexible and pragmatic but also considerably more humane than the approach taken by Havana's communist counterpart in Pyongyang. Overall, the regime in Havana enlisted the Cuban population in an aggressive import substitution program. The policy was a tall order for a country that continued to suffer from the historical trade embargo imposed by its most obvious economic partner, the United States. As a consequence, tractors had to be substituted with oxen, and fertilizer with manure, in order to revitalize agricultural production and feed the population.

At any rate, the real miracle was performed by the Cuban people. Against all odds, ordinary people managed to get by due to the remarkable cohesion of Cuban society at the level of local communities and neighborhoods. Although Cuba is highly urbanized, the typical *barrio* is an urban village. Cuba's multigenerational family households are tightly embedded in neighborhood life. The typical household is shared by an extended family including aunts, uncles, and cousins. One-person households are very rare. Most families have lived in the same home for generations. The occupational structure tends to be mixed, with some members of a household working in the official sector, others in the informal economy, and yet others dedicated to reproduction and care. People cultivate close relationships with friends and relatives inside and outside their *barrio* (Taylor 2009; cf. Rosendahl 1997; Jatar-Hausmann 1999).

One should not idealize this. In the early 1990s, families were stuck in their homes because the regime had frozen the property structure after the revolution. Thus, people were cramped into narrow spaces because they had no other choice. The regime had invested in community cohesion not so much to create social glue, but rather to sustain political control. Moreover, communitarian neighborhood life is not just cozy. It is also rife with gossip and strife (Pertierra 2011; Lewis, Lewis, and Rigdon 1978).

Be that as it may, what ultimately matters is that most Cubans could rely on their families, friends, and neighbors. In a survey, 86 percent of people from vulnerable neighborhoods in Havana declared that they could count on support from relatives, 97 percent from friends, and 89 percent from neighbors (Taylor 2009, 142). This local solidarity, or social capital, helped ordinary Cubans to make ends meet during the Special Period. As one inhabitant of a vulnerable neighborhood put it, the crisis brought people closer together because it forced them to rely on one another (as quoted in Taylor 2009, 140).

In the countryside, there were deliberate efforts to link people with the land. Labor organization on state farms was shifted from collectivist "brigades" to the territorial organization of workforce by ranches (*granjas*) and farms (*fincas*), which were further subdivided into dairies (*vaquerías*) and plots (*lotes*). State farms and agricultural cooperatives were expected to provide their own food, both for canteens and for private consumption. Some factories had workers cultivate land to cater to their food needs. Elsewhere, workers were encouraged to have their own small plots where they could produce food for their families. Thus, localities in the Cuban countryside became increasingly self-sufficient (Deere, Pérez, and González 1994).

Traditional knowledge was another decisive factor in feeding the population. Although most land had been collectivized after the revolution of 1959, about 4 percent of Cuban farmers had kept their land. Another 11 percent was organized in private cooperatives (Burchardt 2000). The survival of traditional family farms and private cooperatives alongside industrial agriculture turned out to be an important asset. Independent farms were more resilient to the crisis than state farms because they operated with less fuel and agrochemical inputs. Cuba's surviving family farmers kept alive important traditional knowledge that could now be recovered. Other formerly independent farmers had moved to towns and cities, where they provided valuable know-how for urban agriculture.

Urban agriculture was a local self-help movement, facilitated by the availability of traditional knowledge in combination with technologies of organic gardening and the Cuban-specific rustic ingenuity. Idle stretches of land between concrete blocks or in urban peripheries were turned into makeshift organic gardens. Vacant or abandoned plots in close vicinity to people's homes were transformed into plantation sites. People used whatever urban wastelands they could occupy to grow vegetables and other foodstuffs.

The movement was purposefully augmented by the regime, but the real action was at the grassroots level. By the mid-1990s, there were hundreds of registered horticultural clubs in Havana alone. An urban cultivator from Havana explained: "When the Special Period started, horticultural clubs were organized by farmers themselves. . . . Special emphasis was made to involve the whole family in these activities. . . . We wanted also to develop more collaboration and mutual help among ourselves; we exchanged seeds, varieties, and experiences. We achieved a sense and spirit of mutual help, solidarity, and we learned about agricultural production" (as quoted in Carrasco, Acker, and Grieshop 2003, 98).

Again, one should not idealize this. Environmentalists have exalted urban farming during the Special Period as a social experiment, or even as an alternative model of organic agriculture.[11] In reality, Cuba's detour into low-input agriculture was obviously driven not so much by ecological consciousness as by dire necessity.[12] From the second half of the 1990s, when the economic situation improved and agrochemical inputs became more available again, many reforms were aborted, and Cuba started drifting back to industrial farming (FAO 2003; Mesa-Lago and Pérez-López 2005). This was helped by subsidized oil deliveries from Venezuela. At the same time, foreign investment enabled Cuba to cover about half of its oil and gas consumption from domestic sources (Economist Intelligence Unit 2008, 24–25).

Nevertheless, it is highly encouraging to note that, during the early and mid-1990s, Cubans managed for a few years to mitigate an extremely disruptive energy scarcity by their remarkable community ethos. The comparison with North Korea shows that this was not a minor achievement.

Peak Oil Trajectories

The cases of Japan, North Korea, and Cuba suggest three different patterns of how different countries or societies may react to a disruptive energy scarcity. Despite the fact that peak oil would initially be experienced as a global energy crunch rather than as a series of national crises, it seems reasonable to expect a comparable gamut of reactions in different parts of the world.

My cases suggest three possible peak oil trajectories, or pathways, that different parts of the world may take in case of disruptive energy scarcity (table 4.2). This obviously does not imply that responses to a terminal decline of world oil production would follow exactly the same lines as the national reactions to oil supply disruptions described in my case studies. Japan in the 1930s, as well as North Korea and Cuba in the 1990s, were unique places. It clearly makes a difference that today all oil-importing countries are tightly integrated in global market structures. Another difference is that, while even a gradual decline of world oil

Table 4.2
Peak Oil Trajectories

Case	Reactive pattern	Outcome
Japan	Predatory militarism	Radicalized imperialism
North Korea	Totalitarian retrenchment	Deindustrialization
Cuba	Socioeconomic adaptation	Communitarian revival

supply would be extremely disruptive for oil importing countries, the onset of mounting energy scarcity after peak oil would be somewhat less abrupt than it was in the cases of North Korea and Cuba.

With all due caveats, the cases discussed offer the best available proxies for the likely social and political effects of peak oil. They are sufficiently analogous to the situation at hand to conjure up scenarios on how different parts of the world would react to a disruptive energy scarcity triggered by peak oil. Countries prone to military solutions may follow a Japanese-style strategy of predatory militarism. Countries with a recent authoritarian tradition may follow a North Korean path of totalitarian retrenchment. Countries with a strong community ethos may be able to embark on Cuban-style socioeconomic adaptation.

It is of course possible to imagine additional reactive patterns, such as the mobilization of national sentiment by populist regimes. Even so, the trajectories identified can help us to derive plausible hypotheses on how different parts of the world would be likely to react to disruptive energy scarcity after peak oil.

Given its unrivaled military capabilities, the United States is the most obvious candidate for a Japanese-style strategy of predatory militarism. Simply put, the United States may be tempted to use its unique power-projection capacity to secure privileged access to oil. It has happened sometimes in the past, and may happen more often in the future, that US decision makers find military coercion more effective than trade. Increased domestic production of unconventional oil in the United States and Canada may obviate the drive for predatory militarism, but this is premised on the continuation of the current boom of shale oil and tar sands. The People's Republic of China (PRC) may be tempted to use its military muscle to secure access to oil and gas in Central Asia and, possibly, in the South China Sea. Elsewhere, the PRC would be unlikely to use a predatory strategy because, for the foreseeable future, its maritime forces and air power are no match for the United States. Countries like India and Israel have even more limited military capabilities, but may nevertheless be tempted to engage in geopolitical operations in their regional neighborhood to secure access to vital energy resources.[13]

A North Korean-style solution of totalitarian retrenchment that screws the population to preserve elite privileges is most likely to occur in countries with a strong authoritarian tradition. In consolidated democracies, totalitarian retrenchment is much harder to imagine. Nevertheless, the history of twentieth-century Europe shows that even liberal democracies can and do sometimes degenerate into tyranny. It is difficult to predict to what point even in consolidated liberal democracies the political culture could deteriorate in a protracted and serious crisis (Gurr 1985). Political elites in less consolidated democracies might experience fewer constraints and scruples right from the start. For example, elites in the second-wave democracies of Latin America may have lesser qualms than their counterparts in Western Europe about screwing their own populace to preserve their privileges.

Compared to predatory militarism and totalitarian retrenchment, Cuban-style socioeconomic adaptation is normatively more desirable. At the local level, people in many developing countries may be able to mitigate the effects of disruptive energy scarcity by reverting to community-based values and a subsistence lifestyle. Such a regression would be comparatively easy for people in societies where individualism, industrialism, and mass consumerism have not yet struck deep roots. By contrast, socioeconomic adaptation would be far more difficult for people in Western societies where individualism, industrialism, and mass consumerism have held sway for such a long time that a smooth regression is hard to imagine. And yet, survival in many presently industrial Western societies may ultimately depend on the mobilization of support from local communities and a subsistence-based lifestyle (Hopkins 2008; Murphy 2008; Orlov 2008; Holmgren 2009; Rubin 2009; De Young and Princen 2012).

In abstract terms, this leaves us with three causal propositions, or hypotheses.

Hypothesis 1   The greater a country's military potential and the stronger the perception that force is more effective than the free market to protect access to vital resources, the more likely there will be a strategy of predatory militarism.

Hypothesis 2   The shorter the time and the less a country or society has practiced humanism, pluralism, and liberal democracy, the more likely its elites will be willing and able to impose a policy of totalitarian retrenchment.

Hypothesis 3   The shorter the time and the less a country or society has been exposed to individualism, industrialism, and mass consumerism, the more likely a regression to community-based values and a subsistence lifestyle.

This is of course not to deny that oil-exporting countries are in a more comfortable position. Other things being equal, they could use the higher revenue from oil exportation to increase their power and wealth, while subsidizing domestic consumption and bolstering their economies—if, that is, they do not fall prey to predatory militarism; and if they evade the so-called resource curse that has bedeviled so many developing countries in the twentieth century, leading to corruption and cementing under-development (Sachs and Warner 1995, 2001).

In the transition, large private Western oil companies such as Exxon and Shell would lose further ground to the state-controlled companies of oil-exporting countries such as Saudi Aramco or Nigeria's NNPC. As a consequence, even oil-importing countries would increasingly rely on state-controlled companies such as China's CNPC (Vivoda 2009). Both in the realm of power politics and the "marketplace of ideas," the ability of Western countries to impose liberal democracy through instruments such as development assistance and economic conditionality would further dwindle (Leder and Shapiro 2008).

This can be formulated as yet another causal proposition, or hypothesis.

Hypothesis 4   In the event of peak oil, there will be winners and losers. It seems reasonable to expect a redistribution of power and wealth from oil importers to oil exporters, and from private to state-controlled companies.

Energy Transition?

So-called techno-optimists object to "peak oil pessimism" that scarcity would not only lead to higher oil prices, but would also trigger a transition to alternate energy sources, such as renewable energy or a new generation of improved nuclear reactors (Smil 2008b). Some optimists even envisage that revolutionary technologies such as solar energy or nuclear fusion could make oil and other fossil fuels entirely redundant (Bradford 2006; Weiss and Bonvillian 2009; Jacobson and Delucchi 2011; Delucchi and Jacobson 2011; but see Trainer 2012).

Could there not be a smooth transition from oil to some alternate resource and technology? Since energy shifts have happened in the past, for example from coal to oil, is it not unimaginative and unnecessarily defeatist to discard such a possibility for the future (Podobnik 2006)? Could there not be a revolutionary technological breakthrough or some other positive surprise around the corner to catapult industrial society beyond oil (or even beyond carbon)?

Alas, time is a serious concern. Developing and rolling out new technologies takes a lot of time. While it is alluring to imagine the sudden appearance of a deus ex machina, such as the discovery of a new energy resource or a revolutionary technological breakthrough, it is important to note that past transitions, including the energy shift from coal to oil, have taken several generations; and that oil was a technically and economically more appealing surrogate for coal; but no such surrogate for oil seems to be available today (Smil 2010; Haberl et al. 2011; Höök et al. 2012; Fouquet and Pearson 2012; Allen 2012; Pearson and Foxon 2012).

Thus, the necessary energy transition beyond oil or beyond carbon is likely to be even more challenging and protracted than past energy shifts. Rather than studying past energy upgrades, we must look at a situation where the challenge was to radically alter an entrenched socioeconomic way of life. This suggests another case study: the US South or Dixieland, after the American Civil War (1861–1865).

What can be gleaned from this case is that the formation of the "new consciousness" necessary for radical social change is a slow and painful process. The socioeconomic backbone resource of the Old South was neither coal nor oil, but human slaves (Mouhot 2011). Precisely because the slave economy worked, white Southerners were willing to defend it in a bloody civil war (Fogel 1989; Wright 2006). After the end of the American Civil War, the forceful abolition of slavery plunged the Old South into a deep crisis.

The Civil War was followed by the Reconstruction Era (1865–1877), when the victorious North tried to enlist dissident elites and former slaves to impose its political and socioeconomic institutions on a reluctant South. Despite the introduction of representation and suffrage for former slaves, reconstruction was mostly thwarted by the recalcitrance of traditionalist Southern elites. Heavy subsidization of railroads by Republican state governments in the South did not lead to the hoped-for modernization, but rather to corruption, making a few investors rich and otherwise contributing to soaring public deficits. After the withdrawal of the last federal troops from the South, race inequality was reestablished under the banner of white supremacy (Fitzgerald 2007).

Later in the nineteenth century, Southern elites started to try to move on. Despite their conservative values, they were not entirely prevented by these from cautiously embracing industrial capitalism. Initially, this amounted to an uneasy compromise between cherished industrialization and dreaded modernization. On the one hand, Southern elites became obsessed with the idea that an industrializing "New South" would rise like a phoenix from the ashes of the "Old South." On the other hand, they remained loyal to time-honored values of agrarianism and patriarchal society. As Mark Twain eloquently put it in 1883, cultural life on the Mississippi was characterized by "practical, common-sense, progressive ideas, and progressive works, mixed up with the duel, the inflated speech, and the jejune romanticism of an absurd past that is dead, and out of charity ought to be buried" (Twain 2006, 264).

This was reflected in the establishment of a quasi-colonial economy. While railroads were finally built on a massive scale, often with capital from the North, industrialization in the South was initially dominated by low-wage and labor-intensive manufacturing. Most industries were dedicated to the processing of agricultural goods (e.g., in cotton mills) or natural resources (e.g., in blast furnaces). The real industrial takeoff came much later, after several generations of socioeconomic backwardness, and was spurred by the New Deal of the 1930s (electrification) and the war economy of World War II. In the mid-twentieth century, Dixieland finally developed as a growth region and came to be seen as part of the American "Sunbelt" (Cobb 1984; Wright 1986). The Civil Rights Act of 1964 famously put an end to official race segregation in the South, although some race issues remain until the present day.

While this amounts to a decently happy ending, it took a century for the South to recover and catch up. This is remarkable because to understand how a technological and socioeconomic upgrade might look, Southerners only had to look to the North of their own country. There, industrial capitalism with its superior technologies and know-how was unfolding before their very eyes. With the right incentives in place, attracting investment and technology transfers from the North would not have been too difficult. But despite such uniquely favorable conditions for industrial development, the century from the Civil War to the Civil Rights Act is replete with unpleasant memories such as race riots and labor revolts, as well as the Ku Klux Klan and Jim Crow laws.

Dixie is a cautionary tale for those who believe that, after peak oil, there will be a smooth technological upgrade. If—even in the US South despite uniquely favorable circumstances—adaptation took a full century, then a technological upgrade will be even harder under the more challenging circumstances of disruptive energy scarcity after peak oil. This time around, the world would be struggling with an industrial downgrade, rather than an upgrade, as in the case of the US South. Developing new energy technologies is never fast and easy, and even less so in times of crisis. After peak oil, we should therefore expect extremely slow and painful processes of social and technological adjustment that may easily last for a century or more (Haberl et al. 2011).

This can be stated as my last and most general proposition, or hypothesis.

Hypothesis 5 In the event of peak oil, we should not expect either immediate collapse or a smooth transition. People do not give up their lifestyle easily. We should expect painful adaptation processes that may last for a century or more.

While this chapter is focused on energy scarcity caused by resource depletion, at least in principle there is also the possibility of climate-change-induced energy scarcity (see box 4.1 on p. 78). Arguably, however, the social and political consequences of the latter contingency would be broadly similar despite some notable differences in geopolitical terms (box 4.2).

Peak Oil Scenario

Based on the insights and hypotheses gleaned from the case studies, it is now possible to develop a *peak oil scenario* of what may happen in

Box 4.2
Social and Political Effects of Climate-Change-Induced Energy Scarcity

In the entirely hypothetical case that a core group of nations (including the United States and China) strikes a global carbon compact, maybe in response to atrocious natural calamities (see box 4.1 on p. 78), it is safe to assume that there will be other nations refusing to join and curtail their fossil fuel consumption. If the global carbon compact limits international trade in high-carbon fuels such as coal, some outsiders to the compact may undertake military operations to secure access to resources located abroad (predatory militarism). Military campaigns are highly carbon intensive and require advanced industrial capabilities, which would be rapidly dwindling among the signatories to the compact. As a consequence, it would become increasingly difficult for the signatories to stop the predatory behavior of the outsiders. It is also safe to assume that the decision to ration fossil fuel consumption would mobilize powerful vested interests in the countries subscribing to the global carbon compact. Elites may secure privileged access to the remaining allocations at the expense of the rest of the population, to the point of establishing antidemocratic political regimes (totalitarian retrenchment). In places where there is significant social capital, ordinary people may be able to discover low-carbon lifestyles based on social solidarity and localized production (socioeconomic adaptation). In other places the descent from industrial society to a new equilibrium may be much harder and last for generations.

different parts of the world during the first couple of decades of mounting energy scarcity.

Please note that a scenario is not a prediction. A scenario is based on *assumptions*. As mentioned above, I follow the peak oil literature in assuming a decline of world oil production by 2–5 percent per year, after a few years on a bumpy plateau. Moreover, I assume that no adequate alternate resource and technology will be available to replace oil as the energy backbone of industrial society.

Three more provisos are in order. First, the scenario provided in this section is limited to the first couple of decades after peak oil.[14] Second, my scenario is deliberately broad-brush and does not offer much detail. Third, nothing of what I am going to say must be understood in a deterministic way. What I can offer is a set of plausible conjectures, rather than scientifically exact point predictions.[15]

Apart from the five hypotheses outlined above, I rely on prior knowledge about historical and institutional path dependencies. While the future is fundamentally open, specific countries and world regions are set on a path that makes some trajectories far more likely than others. For example, we roughly know which countries have strong power projection capabilities, recent authoritarian traditions, and high social capital. We also know which regions possess significant reserves of fuel resources, and how these resources have been managed so far. All of that knowledge can be used for scenario building.

In *North America*, the United States combines strong dependency on foreign oil deliveries with an unrivaled capacity to project power. The current surge of shale oil may postpone this for a decade or two, but ultimately a military strategy will be tempting. To be sure, America's liberal democracy and free-trade ideology militates against the open recourse to military coercion. The United States is going to support liberal democracy and the free market for oil as long as it is convenient. Even when the oil market comes under pressure because of tightening international supply, the United States is likely to continue to defend it for a while. But when soaring oil prices start crippling the American economy, US leaders may find that military coercion is more effective and can be justified in terms of protecting free trade. The United States is then likely to put the blame on foreigners and pursue a geopolitical strategy of energy security to protect the free market and/or the American way of life (Klare 2004, 2008).[16] Why keep negotiating with recalcitrant leaders such as Hugo Chavez if there is a military option? This is not to say that the military option is easy, as the Iraq War has shown. Moreover, liberal democracy in the homeland can be corroded by illiberal practices abroad. Nevertheless, military coercion is likely to gain ascendancy relative to free-market rhetoric as oil supplies dwindle. The resource-rich neighbors of the United States, Canada and Mexico, are likely to be tied more closely to the US core.

In *Latin America*, medium-sized oil-exporting countries such as Venezuela and Ecuador may try to profiteer from soaring oil prices. If they engage in a strategy of brinkmanship and deny the United States oil on favorable terms, then their political regimes may be toppled. While this would further increase anti-American resentment in the region, political elites are likely to acquiesce, ultimately, to US hardball tactics. In fact, historical evidence suggests that Latin American elites often opportunistically collude with the United States. Eventually, resource-rich Brazil may be able to escape intervention due to its larger size and geographical distance from the United States. If Brazil manages to offer sufficient benefits to neighboring countries, a regional state complex around Brazil may eventually be possible. Otherwise, energy-poor Latin American countries would enter a serious crisis. We may then see how much Cuban-style socioeconomic adaptation is possible in other Latin American societies.

Western Europe would enter a particularly difficult quandary. In theory, advanced industrial countries such as Germany and France could quickly rearm. In practice, however, predatory militarism is not a credible option for them. Since Europeans have good historical reasons to dread militarism, the social consensus necessary for this strategy would not be forthcoming at the decisive initial stages of geopolitical positioning. For the same historical reasons, in most of Western Europe the path of totalitarian retrenchment does not seem to be available either. Concomitantly, Western European countries would be forced to strike opportunistic bargains with Russia and oil-exporting countries across the Mediterranean. Due to their asymmetrical nature, such deals would be inherently fragile and subject to constant renegotiation. Investment in renewable energy and innovative technologies could somewhat smooth the transition, but ultimately Europeans would hardly be able to avoid a transition to a more community-based lifestyle. Despite the present affluence of Western European societies (and, in part, precisely because of it), this would be extremely painful and last for generations.[17]

Ordinary Western Europeans would be forced to rely on local communities for their welfare, if not their survival. For most of the indigenous population, a regression to community-based values and a subsistence lifestyle would be challenging because the habits of industrial society are deeply rooted. The problems would be compounded by the fact that immigrant groups might segregate from Europe's multiethnic societies, potentially reinforcing religious fault lines. On the one hand, this might enhance the solidarity among members of specific social groups. On the other hand, it would almost certainly conjure up severe social conflict.

The situation in *Japan* would be largely comparable, although Japan is far less multiethnic and Japanese people may be more willing to accept disruptions to their taken-for-granted lifestyles. This was confirmed in 2011 when the Japanese responded in a calm and disciplined way to a tsunami followed by serious mayhem and a nuclear meltdown at the facilities in Fukushima. As in the Western European case, however, the unavoidable transition to community-based values and a subsistence lifestyle would be painful and last for generations.

The situation would be somewhat different in countries and regions that have industrialized later and/or have a more recent authoritarian tradition that can be recovered. Therefore, totalitarian retrenchment and socioeconomic adaptation are more likely and easier to imagine in the new democracies and semi-authoritarian countries of *Eastern Europe* and *Southeast Asia* than in Western Europe, Japan, Australia, Canada, or the United States.

In *least developed countries (LDCs)*, common people with limited exposure to industrial lifestyles would be forced to rely on the cohesion of social groups for their survival. Given the high population pressure in most LDCs, however, large population segments would fall victim to famine, disease, and conflict. Particularly but not exclusively in *sub-Saharan Africa*, state failure and conflict over scarce resources would become endemic. Moreover, the inevitable end of the oil-based green revolution in agriculture and the demise of international aid would wreak environmental havoc and human insecurity. The production of biofuels and other cash crops might improve the situation of wealthy strata, but would crowd out food production and thus exacerbate the plight of the poor. The ecological situation would be aggravated by vital biomass being removed from the soil as a combustible. In most places, the unavoidable consequence would be famine, disease, and mass exodus. In some places, however, a revival of community-based values and a return to a subsistence lifestyle might mitigate the effects (Shiva 2008).

The elites of oil-exporting African kleptocracies such as Nigeria, Angola, and Equatorial Guinea would certainly keep selling their oil to the highest bidder, especially when the bid is backed by sufficient military clout, and when there are no onerous obligations with regard to democratization and human rights. If the United States gives up its dysfunctional democratization agenda, it will have better access to African resources than Europe, China, or Japan. However, it is an open question how much ordinary people in African petro-states would benefit from the increased oil revenues (people in African countries that do not have rich fossil fuel reserves would almost certainly suffer more).

In Asia, *Russia* has enough resources to provide for its own energy needs. In geopolitical terms, it would become a more important regional player due to its abundant export capacities. *China*, by contrast, heavily relies on imported oil.[18] To preserve its industrial capacity, the country might be tempted to secure access to vital resources from Central Asia by military means. Totalitarian retrenchment may also be lurking. *India* has more limited military clout and a less authoritarian state tradition, but may nevertheless be tempted to engage in limited geopolitical operations in its regional neighborhood. Small and resource-poor outposts of industrial civilization, such as Singapore, would struggle to survive.

The oil-exporting countries of *Central Asia* and the *Middle East* would benefit more than in the past from their abundant resource endowment. Due to the effects of skyrocketing oil prices on the world market, their economies would continue to grow in relative and absolute terms. Their domestic oil consumption would be stable or even increase at a time when it would be declining in the rest of the world (Rubin 2009, 57–83). While the "resource curse" would persist in countries with particularly corrupt elites (Sachs and Warner 1995, 2001), in others political freedom may improve alongside the level of industrialization and the standard of living (Dunning 2008). The Middle East would almost certainly replace Western Europe as the most attractive destination for Muslim migrants.

Bottom Line

This is not a cozy world to imagine. Most of us would clearly prefer industrial civilization to continue unabated, perhaps somewhat mitigated to avoid the worst effects of anthropogenic climate change. But even though we may not like the prospect of disruptive energy scarcity, it would be utterly imprudent not to seriously consider its social and political implications.

5

The Struggle over Knowledge

A battle is raging over knowledge about climate change and energy scarcity. The fundamental bone of contention is the same in either case. Do we really have to worry, or can we discard it all as scaremongering? In either case, the answer to that question determines the frontlines of the epistemic battle.

In the case of energy, the stronghold of mainstream expertise is kept by those who argue that resource issues are no serious reason for concern because they are taken care of by the market mechanism and technological progress. Small but vocal groups of alarmists besiege that stronghold, enraged by the cold refusal of the energy experts to acknowledge the gravity of the situation. The fortress is holding steadfast, despite some cracks appearing in the wall.

In the climate case, the roles are exactly the reverse. Those who believe that climate change is a dreadful problem are firmly entrenched on the hilltop of mainstream science. Down in the valley are gathering those who refuse to see climate change as a serious issue, enraged by the sometimes condescending and sometimes contemptuous attitude of the alarmists. Obviously, the incumbents see no reason to cede the stronghold to their challengers.

There is a puzzle. Climate change and energy scarcity are both fundamental challenges to the viability of industrial civilization. How is it possible that, in the case of climate change, the alarmists have come to represent mainstream science, whereas in the energy case they have never made much headway?

To understand this puzzle, I start off by developing an analytical framework that enables us to explore the struggle over knowledge about energy scarcity and climate change. I outline three kinds of science:

normal, abnormal, and post-normal. I further show that, depending on whether normal or post-normal science reigns supreme, there are different patterns of contestation.

The framework is subsequently applied to energy and climate science. In the energy case, normal science is under siege from the abnormal science about scarcity. In the case of climate change, post-normal science is caught in a stranglehold, or double whammy, between two different breeds of abnormal scientists: radical alarmists criticizing the post-normal establishment; and contrarian scientists alienated by the shift from what they would consider normal, "value-free" science.

The narrative account for each knowledge regime is complemented by an Internet-based cluster analysis of the relevant networks, graphically illustrated by the cluster maps in figures 5.3–5.6. All cluster maps have been produced with the free software Issue Crawler and indicate how specified websites are (hyper)linked to others in citation networks.[1]

A systematic comparison between the cases explains the different patterns of contestation in either field and tackles the crucial question of whether and how post-normal science affects the prospects for either problem, or both, to be addressed. Despite the interesting differences between climate science and the study of energy it turns out that, given the equally dismal outcome in either case, scientists are in a double bind: they are damned if they do and damned if they don't engage in post-normal science.

Three Kinds of Science

Thomas Kuhn (1962) famously understood normal science as a social enterprise where an epistemic paradigm is shared by a community of scientists in such a way as to enable incremental research. In normal science there is an agreed framework of what constitutes a problem, as well as what the relevant facts are and how to interpret them. Residual uncertainty is acknowledged but seen as temporary. Scientists do not need to engage in normative debate and advocacy for specific solutions because anything that is overtly political does not belong to the "value-free" sphere of science.

While many scientific fields such as mathematics, metallurgy, or structural engineering are squarely within the remit of normal science, late
modernity increasingly leads to situations where "facts are uncertain,
values in dispute, stakes high and decisions urgent" (Ravetz 2004, 349).
In such post-normal situations, debates over uncertainty go beyond mere
technicalities and include radical doubt and ethical contestation (Funtowicz and Ravetz 1993). Because there is serious dispute around the most
fundamental values to be promoted or defended, the knowledge at stake
seems too existential and too political to be left to the established experts
practicing normal science. As a consequence of such fundamental contestation, normal science is challenged by what, for lack of a better term,
I call abnormal science.

A post-normal situation can lead to an opening in the scientific
establishment, with reformist scientists willing to reconsider the bound-
aries of science while advocating policies to address the issue at hand.
Moderate exponents of abnormal science will then attempt to shape, or
at least influence, the reformist agenda within the scientific establishment
(figure 5.1).

The harmony of interest between the two groups enables an extension
of the scientific peer community to incorporate moderate exponents of
abnormal science such as interested citizens, progressive politicians, and
media pundits. Unlike their more radical fellows, moderate critics natu-
rally welcome the opportunity to be coopted into the extended peer community (Healy 1999). Their cooptation into the establishment then leads
to what is called post-normal science (Funtowicz and Ravetz 1993).

The passage from normal to post-normal science leads to one fusion
and two splits. It leads to a fusion between reformist scientists and
moderate critics and a split between former bedfellows in the critical
camp, as moderates are absorbed into the extended peer community

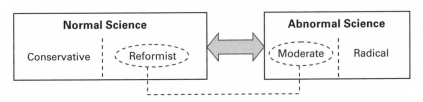

Figure 5.1
Normal and abnormal science

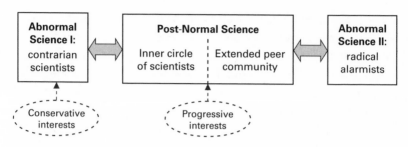

Figure 5.2
Post-normal and abnormal science

while radicals resist cooptation. It also leads to another split between those scientists willing to go along with the move to post-normal science and those objecting to the overt politicization of science. As a result, science is caught between two different strands of abnormal science: radical alarmists resisting co-optation, and contrarian scientists resenting the fact that science is not normal any more (figure 5.2).

The intense polarization inherent in this gambit brings forth a situation where post-normal science receives support from progressive interests pleased by the reformist policy agenda, while conservative interests provide support to disgruntled former normal scientists who fiercely denounce the "unscientific" muddling of research with political commitment.

Take for example the issue of genetically modified maize (De Marchi and Ravetz 1999). Over the 1990s, establishment science was fiercely irritated by the mobilization against "gene maize." Normal scientists felt under significant pressure to extend the peer community and allow motivated outsiders to participate in the scientific debate. In Europe the result has been the cooptation of these outsiders into post-normal science. In the United States and most other parts of the world the pressure was resisted, with the result that normal science continues to be haunted by the same critical voices as before.

Even in "post-normal" Europe, there were radical critics of genetically modified organisms (GMOs) who did not want to get themselves coopted by the new mainstream. Post-normal science has not dissuaded such radical alarmists from opposing, on principle, any form of gene manipulation, including what is deemed reasonable by mainstream scientists. On the opposite end of the political spectrum, alienated former establishment figures have retooled as contrarian scientists contesting the new post-normal science.

As the example shows, abnormal science does not cease after the cooptation of outsiders. No matter how many former critics are coopted, there always remains a "lunatic fringe" of radical alarmists opposing science no matter what accommodation has been reached by the extended peer community. At the same time, the cooptation of outsiders typically leads to strong resentment among previous insiders opposing the post-normal policy agenda.

This is exactly what can be seen in the case of climate change, where science became post-normal early on. And yet this has not led to the disappearance of abnormal science. Radical alarmists denounce the "climate consensus" as lackadaisical, while contrarian scientists posture as climate skeptics blowing the whistle against the same consensus for being unduly alarmist.

The energy case is remarkably different. Like climate change, energy scarcity is easily identified as a post-normal problem. And yet, post-normal science has come to dominate only climate science, not the study of energy. In the energy case, there is a longstanding pattern of normal science challenged by abnormal science but refusing to become post-normal. The result has been an entrenchment of normal science and a radicalization of abnormal science.

To sum up: In a post-normal situation, when science is challenged by dissident outsiders, it sometimes becomes post-normal and sometimes resists. In some cases, such as climate change, post-normal situations lead to the cooptation of moderate outsiders into the mainstream, spawning post-normal science. In other cases, such as energy, normal science becomes entrenched. Either way, radical challengers persist. As long as post-normal problems remain politically salient, the nuisance posed by abnormal science never disappears.

Let us now apply this framework to the specific knowledge regimes and patterns of contestation found in energy studies and climate science.

Energy Studies: Normal Science under Siege

Despite a convincing case that the risk of energy scarcity is a post-normal problem (Tainter, Allen, and Hoekstra 2006), normal science continues to reign supreme in the field of energy studies. Mainstream economics is the science of choice when it comes to making forecasts, despite the fact that more technical disciplines such as geology and engineering are more directly concerned with fuel extraction and, by implication, future energy supply. Official expertise is concentrated in a couple of authoritative national and international institutions: the International Energy Agency (IEA) and its US counterpart, the Energy Information Administration (US-EIA).

Though not very successfully, normal energy expertise is challenged by a critical counterculture of abnormal energy science that is mostly made up of concerned citizens and a small number of retired geologists sounding the alarm at runaway fuel depletion. This abnormal energy science has its own institutional infrastructure, albeit relatively dispersed. At its center, there are loose epistemic networks like the Association for the Study of Peak Oil and Gas (ASPO) and its various national offshoots, as well as a number of Internet platforms such as *The Oil Drum* and *Energy Bulletin* (which moved to resilience.org in January 2013).

Normal Energy Expertise

Energy expertise is concentrated in a small number of organizations gathering and processing data (table 5.1). By far the most authoritative entity is the International Energy Agency (IEA), which publishes the

Table 5.1
Authoritative Sources on Energy Supply

		Key publications	Time horizon
IEA	Established in the framework of the OECD	*World Energy Outlook; Energy Technology Perspectives*	until 2035 until 2050
US-EIA	Analytical agency within the US Department of Energy	*International Energy Outlook*	until 2035
BP	Business firm	*BP Statistical Review BP Energy Outlook*	until 2030
Shell	Business firm	*Shell Energy Scenarios*	until 2050

annual *World Energy Outlook*, as well as a biennial report called *Energy Technology Perspectives*.

The United States Department of Energy's Energy Information Administration (US-EIA) is somewhat less authoritative because it represents a particular government. In recent years, US-EIA has been more optimistic about future energy supply than the IEA. Despite this bias, or perhaps precisely because of it, some observers prefer to quote US-EIA rather than the IEA.

Business firms have limited epistemic authority due to their vested interests. Nevertheless, industry-oriented circles sometimes rely on their publications. Interestingly, a feature common to all energy supply reviews is that data are not directly gathered but collected from sources such as the United States Geological Survey, the OPEC Secretariat, the *Oil and Gas Journal*, or the *World Oil* magazine (Owen, Inderwildi, and King 2010, 4744).

Overall, the Paris-based International Energy Agency is so authoritative that it can be studied as a proxy for the energy establishment more generally (Van de Graaf 2012). Since its foundation in 1974, the IEA has slowly morphed into the "global energy policy advisor" par excellence (Van de Graaf and Lesage 2009, 314). It has become "an agency for compiling data and making forecasts on energy markets" (Noreng 2006, 48), the hallmark of which is the collection of statistical data and the publication of the widely acclaimed *World Energy Outlook* and the *Energy Technology Perspectives*.

None of this was preordained. The original raison d'être of the IEA was to serve as an insurance regime against major oil supply disruptions (Willrich and Conant 1977; Keohane 1984, 217–240; Bamberger 2004). Its mandate was to make wealthy oil-importing countries less vulnerable to supply disruptions such as the oil crisis of 1973–1974. In line with the widespread anxiety of the early 1970s about limits to growth (Meadows et al. 1972), another task of the IEA was to prepare its member states for the risk of chronic undersupply (Van de Graaf and Lesage 2009; Colgan 2009; Kohl 2010).

In the heat of the oil crisis, US secretary of state Henry Kissinger (1973) declared in an address to the Pilgrim Society that "we must bear in mind the deeper causes of the energy crisis: it is not simply a product of the Arab-Israeli war; it is the inevitable consequence of the explosive growth of world-wide demand outrunning the incentives for supply. The Middle East war made a chronic crisis acute, but a crisis was coming in any event. Even when pre-war production levels are resumed, the problem of matching the level of oil that the world produces to the level which it consumes will remain."

On Kissinger's initiative, the International Energy Agency was established in 1974 by oil-importing countries as a counter-cartel to the club of oil-exporting countries, OPEC. According to the bylaws of the IEA, the most immediate task was to establish and manage a crisis response mechanism. Member states were obliged to hold the equivalent of a few months of net oil imports as an emergency stockpile. In case of a major oil supply disruption, defined as a shortfall of oil supply of 7 percent or more, the IEA Secretariat would have the authority to declare an emergency. Member states would then be obliged to share their supplies and implement appropriate demand-restraint measures.[2]

Since its creation the IEA has been largely dormant as an insurance regime. There has not been a single event that would have met the agency's formal definition of a major oil supply disruption: shortfall of oil supply of 7 percent or more (IEA 1974).[3] The IEA has acted only three times, on a voluntary basis, to inject additional oil from stockpiles into the market through coordinated action: in 1991, as a response to the Gulf War; in 2005, in response to Hurricane Katrina; and in 2011, to offset the disruption of oil supplies from Libya.[4]

Although the IEA has never really been tested as an insurance regime, it is a sizeable international bureaucracy. According to its website, the agency has "a staff of 260 enthusiastic professionals" (IEA 2012f). Like a fire department, it must always be on standby in case there is a real emergency.

Precisely because the IEA has been inactive as a fire fighter, its staff have had to be employed in some other useful way. For that purpose, the statutes of the IEA mention a few other goals in addition to the emergency response mechanism. The most important are monitoring the oil market and reducing the dependency on imported oil (IEA 1974; for the details see Van de Graaf and Lesage 2009; Colgan 2009). Thus, the IEA was originally mandated to evolve in two complementary ways: first, to keep track of international markets and thus provide an early warning mechanism; and second, to work on ways to reduce the unsustainable oil dependency of industrial countries.

While the second task might have suggested a shift to post-normal science, things did not turn out that way. The agency has eagerly embraced the first task of monitoring international markets, which was entirely in line with normal science. However, it never really confronted the task of questioning the oil dependency of industrial society. This was a regrettable failure of policy foresight. Insofar as one of the IEA's core missions was the strategic governance of energy scarcity as a long-term risk, it would have been mandatory for the agency to develop into an expert watchdog to comprehensively monitor the availability of oil and to facilitate a large-scale transformation away from oil and other non-renewable energy resources. But this did not happen for a number of reasons.

First, the IEA was never really meant to question the presumption that oil is abundant. By placing its faith in markets, the agency followed the policy preferences of its member states. It is true that, since the 1990s, some members have been busy "greening" the IEA (Kohl 2010). But no country has seriously urged the IEA to investigate the risk of energy scarcity.

Second, the decline of oil prices from the mid-1980s made the task of preparing industrial society for the eventuality of disruptive energy scarcities appear less urgent and allowed the IEA to focus on standard operating procedures like gathering data, developing forecasting tools, and publishing at the end of every calendar year the iconic *World Energy Outlook*.

Third, the IEA is formally attached to the Paris-based club of industrialized countries, the Organisation for Economic Co-operation and Development (OECD).[5] The institutional culture of the OECD has always been characterized by a firm belief in the capacity of markets to safeguard economic development. It is easy to see that the ideological commitment of the OECD to standard economic thinking has rubbed off on the IEA.

Fourth (and closely related), for a long time the IEA has been dominated by mainstream economists. As in the OECD, most staff members are economists and/or public servants, usually with a background in economics. There have always been a few lawyers, but engineers, geologists, and other energy experts have been a small minority until recently. The long-standing ascendancy of mainstream economists has been consequential. For most economists it is simply axiomatic that, in an effectively functioning market, supply will always meet demand. Accordingly, until 2008 the standard practice of the IEA has been to extrapolate trends in energy demand, and simply to assume that future demand will be met via the market mechanism.

The result is paradoxical. Originally, the whole point of setting up the IEA was that oil supply cannot be taken for granted. Oil markets can be disrupted, not only for political reasons but also by physical scarcity. It would have been a small step from trying to manage geopolitical risks such as the 1973 oil crisis to considering the geological limits to the future supply of oil and other vital energy resources. Paradoxically, however, the idea of limited resources became anathema to the very watchdog of oil supply disruptions.

In the 1998 *World Energy Outlook*, and then again in the 2008 *WEO*, the IEA (1998, 2008) looked more carefully into the physical availability of energy resources. In both cases it appears that in subsequent years there was backlash from member state principals and particularly the United States (Macalister 2009; Badal 2010). Presumably as a result of such backlash, the IEA has become more optimistic again (IEA 2009, 2010a; criticized by Miller 2011). The latest edition of the *WEO* (2012a) is strikingly upbeat compared to previous iterations, despite the fact that only few fundamentals have changed. But even in the notoriously "alarmist" 2008 *WEO*, the IEA made highly debatable assumptions about decline rates and supply following demand (Aleklett et al. 2010).

In principle, the risk of energy scarcity would have been a candidate for post-normal science. Oil depletion clearly is a field where "facts are uncertain, values in dispute, stakes high and decisions urgent" (Ravetz 2004, 349). Data on reserves are notoriously uncertain; insofar as the industrial way of life is at stake, energy is deeply intertwined with fundamental values; the stakes are enormously high; and appropriate decisions are, or would be, very urgent. Despite the fact that all of these conditions are met (see Tainter, Allen, and Hoekstra 2006), the study of oil depletion has not become a post-normal science. On the contrary, it is largely synonymous with the establishment perspective, as can be gleaned from an Internet-based cluster map (figure 5.3).

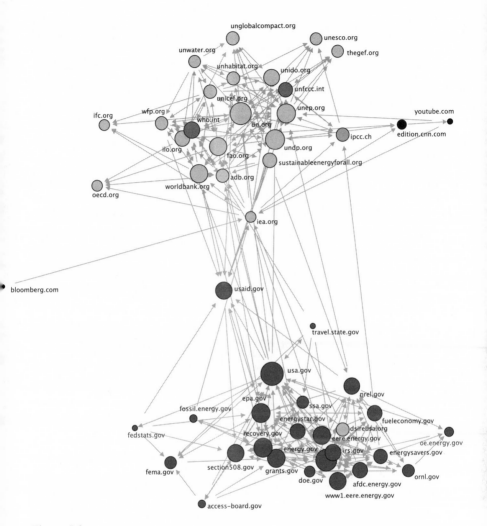

Figure 5.3
Internet-based cluster map of normal energy expertise. Analysis conducted on
March 16, 2012, with issuecrawler.net. Starting points: iea.org; worldenergyoutlook
.org; eia.gov; energy.gov; worldenergy.org.

Three features are remarkable. First, there are two discrete network clusters, with the IEA embedded in one and US-EIA embedded in the other. Second, the two clusters have only few links to each other. Third, a qualitative examination of the top fifty nodes in the network reveals that only fifteen are related to energy, and only five have anything to do with science. Instead, most of the linkages are either between governmental bodies or between international agencies. There are only few linkages to the non-official world.

The US-EIA cluster is exclusively composed of government websites, to the degree that US-EIA does not even appear on the cluster map because it is crowded out by more intense linkages among other network participants. Similarly, the IEA cluster is mostly composed of international agencies. The only exceptions are three linkages between international agencies and non-official sites.[6] All three of them are with the media, and there are no linkages at all with the energy industry or outside critics such as "peak oilers."

Far from supporting the myth that normal science is apolitical, the network is entirely dominated by the political establishment.[7] While there is no way around the IEA and, to a lesser degree, US-EIA, these agencies hardly relate to anyone outside the official world of public policy making.

Abnormal Energy Science

The study of energy scarcity in general and peak oil in particular is a paradigm case of the abnormal science of radical dissidents.[8] It is civic and intellectual alarmism gone wild in the face of a serious post-normal problem, and in the absence of a willingness on the part of normal science to reconsider its tenets, extend the peer community, and thus become post-normal.

Abnormal energy science is conventionally traced back to the founding father of peak oil theory, Marion King Hubbert. Like many other oil men, this leading geoscientist was a self-made man and a maverick. Unlike most of his colleagues working for the oil industry, however, he worried about physical limits to growth. Hubbert gave a first warning of oil depletion in the 1950s (Hubbert 1956), long before the iconic *Limits to Growth* study (Meadows et al. 1972), and he kept emphasizing the physical limits to growth until his death in 1989 (and even posthumously, in Hubbert 1993). This made him an outsider both among his colleagues at the Shell Research Laboratory in Houston, Texas, and in his academic peer group of geoscientists and oil geologists.

Ever since then, the most prominent proponents of peak oil theory typically have a background in geology, engineering, or some other physical science. However, their dissident standpoint forces them to turn their back on normal science and seek an audience among outsiders, notably concerned citizens.

During the 1980s the debate was largely dormant due to low oil prices, but it was reignited in the early 1990s through a book written by geologist Colin Campbell as he neared his retirement (Campbell 1991). A few years later, Campbell (1997) published another book about oil depletion that greatly benefitted from data provided by the company Petroconsultants. The following year, he partnered with retired petroleum engineer Jean Laherrère to publish an article in *Scientific American* (Campbell and Laherrère 1998), which is often cited as the beginning of the contemporary peak oil debate. Their research received tacit recognition in the 1998 *World Energy Outlook* (IEA 1998). Subsequently the IEA officially returned to its sanguine view of oil reserves, but the notion of peak oil was further popularized in a book by retired oil geologist Kenneth Deffeyes (2001).

Forty years after Hubbert's original analysis (1956), and despite the fact that the oil shocks of the 1970s were rapidly fading from memory with prices heading toward record lows, authors such as Campbell and Laherrère were trying, from their retirement, to awaken the world to what they saw as the defining challenge of the twenty-first century. Needless to say they were mostly ignored and sometimes opposed by their mainstream colleagues, with particularly fierce criticism and even ridicule coming from economists and industry figures unwilling to accept the idea of oil depletion (Adelman 1995, 2004; Odell 2004; Maugeri 2006, 2012; Clarke 2007).[9]

Although public interest in oil depletion was initially limited to the "lunatic fringe," physics professor Kjell Aleklett at the University of Uppsala in Sweden organized a conference in 2002 and used it to shepherd Campbell and the other members of the fractious peak oil community into ASPO, the Association for the Study of Peak Oil (Campbell 2011; Bentley 2011). Despite the lack of funding, this independent and loose collection of individuals, mostly retired geologists and academics from a broad range of fields, has since played an important role as an institutional platform, convening annual congresses on peak oil and coordinating various national chapters.

This became possible because, from about 2003, increasing oil prices and the surge of Web 2.0 brought an explosion of peak oil citizen science. The takeoff was further catalyzed by Richard Heinberg's influential and popular book *The Party's Over* (Heinberg 2003), as well as the peak of North Sea oil and gas and the outbreak of the Iraq War, which was often described as a war for oil.

The isolationist radicalism of the peak oil counterculture was particularly evident in the blogosphere, where sites such as *Energy Bulletin* (from 2003), peakoil.com (from 2004), and *The Oil Drum* (from 2005) took off rapidly and gained additional speed in 2005, when Hurricane Katrina shocked the United States, taking offline a significant portion of oil production and refining capacity. As the founder of *Energy Bulletin* recalls:

The issue of energy depletion had very little representation on the web . . . in 2003 when I began work on *Energy Bulletin*. The most prominent site about peak oil was Jay Hanson's DieOff.com with its animations of grotesquely obese Americans overlaid with dead bodies and famine. . . . It was collapse porn. . . . Engaging with the issue felt like stepping into an alternative reality, and quite a lonely one! . . . I chose the rather generic name *Energy Bulletin*, and neutral color scheme to suggest a certain amount of "neutrality," and perhaps to obscure the fact that myself and my colleagues were working on it in our spare time from our bedrooms and secretly from our workplaces, and had no formal qualifications in the areas of either journalism or energy.[10]

Julian Darley, Web host and founder of the platform *Global Public Media*, points out that niches and cliques are common in the counterculture, and that those worrying about peak oil have "interesting personalities that make them not very acceptable to the mainstream. A lot of people do it because it is different. The moment people start agreeing with them, they become fractious, start arguing with each other." He adds that "some counterculture enthusiasts even like being awkward, spiky. You have the wrong people making the wrong kind of arguments, even though they might have the facts right."[11]

At the more scholarly blog *The Oil Drum*, the association of the peak oil community with abnormal science is equally apparent. The site was founded by Kyle Saunders, professor in political science at Colorado

State University, and David Summers, professor of mining engineering at the University of Missouri-Rolla. Both initially wrote under pseudonyms. Stories of academics losing tenure-track positions because of a blog were legendary in the blogosphere, which is why "Prof. Goose," aka Kyle Saunders, would not risk revealing his name until getting tenure (McKenna 2007, 224).

All peak oilers share at least three views in common: that oil is a finite resource, that it is essential to industrial civilization, and that its production peak is fairly imminent. As one insider puts it, the issue of peak oil is "at once frightening, invigorating, and often strangely unifying: you might find a remarkable many agreements in the views of a conservative politician, anarcho-primitivist or ach-druid who write on the topic."[12]

Another remarkable common feature is that most on- and offline forums discussing peak oil are strongly male dominated. A 2009 readership survey carried out at *The Oil Drum* found that more than 90 percent of respondents were male.[13] Two years later the 2011 ASPO conference held in Brussels had 217 participants, 82 percent of whom were male.[14] On the now-defunct discussion forum *Life after the Oil Crash*, this was rationalized in terms of women having difficulty facing the truth because of attachment to family and children. In this context, it is worth noting that the community activists of the more "positive" Transition Town movement (Hopkins 2008) are characterized by a significant departure from the usual male over-representation, if anything reversing it.[15]

Overall, and despite the commonalities, there are three different camps in the community. The first camp emphasizes the limits to growth and sees oil depletion as just one limitation alongside environmental sinks reaching their capacities, ecosystems being exploited, and the depletion of other finite resources. Rather than a problem in and by itself, peak oil is seen as a symptom of a wider malaise facing growth-based complex societies (see already Hubbert 1993). In this view, "culture is sleepwalking" toward "the end of growth," and peak oilers are "rooting for cultural change."[16]

Another camp sees peak oil as a liquid fuels problem. This view is particularly prevalent in the United States, where it was promoted by the influential *Hirsch Report*. Commissioned by the US Department of Energy, the report's lead author, Robert Hirsch, describes peak oil as "an unprecedented risk management problem. As peaking is approached, liquid fuel prices and price volatility will increase dramatically, and, without timely mitigation, the economic, social, and political costs will be unprecedented. Viable mitigation options exist on both the supply and demand sides, but to have substantial impact they must be initiated more than a decade in advance of peaking" (Hirsch, Bezdek, and Wendling 2005, 4). Framed this way, peak oil represents an enormous challenge, but in case of timely action the problem need not be fundamental.

The third group is best described as "doomers," with many happily labeling themselves that way. In their view, peak oil represents a desperate problem without any hope of significant mitigation, and it will inevitably lead to a partial if not total collapse of civilization as we know it. A leading voice in that area was Matt Savinar through his website and online forum *Life after the Oil Crash*, founded in 2003. Savinar abruptly closed the site in 2011, splitting the doomer community across several spinoff websites.

An Internet-based cluster analysis of the peak oil community (figure 5.4) reveals that here are a few "spiders in the web" including ASPO's homepage peakoil.net, *The Oil Drum*, and *Energy Bulletin*. Despite the presence of these hubs, the network is highly dispersed. It is clearly more pluralistic and inclusive than its counterpart, the official network surrounding the IEA and US-EIA (figure 5.3). Nevertheless, peak oilers mostly surround themselves with like-minded critical outsiders or mavericks from the establishment, as would be expected from a community representing abnormal science in the absence of a shift from normal to post-normal science (figure 5.2).[17]

Another remarkable feature is the one-sided focus on resource depletion, especially of oil. While there are sound theoretical and empirical arguments for peak oil, energy scarcity may equally be triggered by a constraint on the amount of CO_2 emissions that can be released into the atmosphere. But apart from a small number of links to the IPCC, climate change hardly registers in the peak oil community. It is however indirectly reflected in the presence of activist outfits that would like to wean industrial society off fossil fuels, such as the Transition Town movement and the Post Carbon Institute.

About half of the fifty nodes in the network are occupied by websites and blogs dedicated to peak oil, plus another five radical activist websites representing the transition and anti-growth movements. The next largest

Figure 5.4
Internet-based cluster map of the peak oil community. Analysis conducted on March 15, 2012, with issuecrawler.net. Starting points: theoildrum.com; peakoil.net; odac-info.org; aspo-usa.com.

group is print and audiovisual media reporting affirmatively about peak oil. Four international agencies, including the IEA and the IPCC, are also in the network, as well as the US Environmental Protection Agency. While a British parliamentary group and government-sponsored research agency sympathetic with the peak oil message are also part of the network, US-EIA is conspicuous by its absence. This is interesting because the optimism of US-EIA about energy supply is diametrically opposed to peak oil.

What we observe here is a radical bifurcation between official normal science and unofficial abnormal science. On the one hand, the economists at the IEA and US-EIA, in unison with engineers and geologists working for the industry, maintain the trappings of normal science and business as usual. On the other hand, whistleblowers and dissidents fill a parallel world in the blogosphere where polemics and eschatological thinking loom large.

The Study of Climate Change: Post-Normal Science in Action

The case of climate change is strikingly different. While in the case of energy studies we have seen normal science under siege, here we see post-normal science in full swing (Bray and von Storch 1999; Saloranta 2001; Hulme 2009).

Post-normal climate science comprises various disciplines that mostly belong to the natural sciences. By far the most authoritative knowledge institution dealing with climate change is the Intergovernmental Panel on Climate Change (IPCC). As would be expected from a post-normal science, knowledge about climate change is also institutionalized in more inclusive and less formal arrangements such as websites and blogs.

As the model of post-normal science suggests (figure 5.2), climate science is challenged by two different forms of abnormal science. One of them is made up by concerned citizens acting as radical alarmists, whereas the other consists of contrarian scientists acting as climate skeptics. The former group is mostly composed by journalists and public intellectuals writing books and/or posting on websites and blogs. The hotbeds of climate skepticism are conservative think tanks such as the Heartland Institute and, crucially, popular scientific websites and blogs such as wattsupwiththat.com and climateaudit.org.

Post-Normal Climate Science

Until the 1950s climate science was a loose configuration of disaggregated disciplines reaching from meteorology to ocean research and from industrial engineering to plant science. During the 1960s the nascent environmental movement and the related academic trend of interdisciplinary "ecological" assessment prepared the ground for the advent of post-normal science.

From the 1970s onward a combination of concerned scientists and committed outsiders started to see climate change as an existential threat to the environment and called for the integration of the relevant disciplines. As a result, during the 1980s all sympathetic stakeholders joined the hodgepodge of post-normal climate science, very much to the chagrin of disgruntled normal scientists and radical alarmists (see Weart 2008, 2012; Hulme 2009).

Since then, climate scientists have colluded with sympathetic public bodies, whether at the national or at the international level, to translate their expert views into political action. The Intergovernmental Panel on Climate Change, established in 1988, was particularly instrumental to this end (Agrawala 1998; Bolin 2007).[18] Ever since the run-up to the UN Framework Convention on Climate Change (UNFCCC), which was concluded at the 1992 Earth Summit in Rio de Janeiro, the IPCC has acted as a strategic link between the science and the politics of climate change (Tonn 2007; Van den Hove 2007).

Based on scientific expertise, the IPCC has collated four assessment reports (1990, 1995, 2001a, 2007a) that have significantly shaped public debates and international negotiations. With increasing levels of urgency and confidence, the reports have suggested that current levels of greenhouse gas emissions are not sustainable. From this, decision makers have derived prescriptions on what ought to be done to get them back on track to a sustainable level.

The first IPCC report (1990) set the stage for the Earth Summit in Rio. The second report found that "the balance of evidence suggests a discernible human influence on climate change" (IPCC 1995, 22). This emboldened the European Union to push for quantified emission targets in the negotiations leading up to the 1997 Kyoto Protocol. Other key international players such as the United States and Japan were mildly sympathetic at the time but showed themselves less convinced by the scientific consensus.

The third report of 2001 concluded more boldly that "most of the observed warming over the last fifty years is likely to have been due to the increase in greenhouse gas concentrations" (IPCC 2001a, 61). While this did not prevent the US administration of George W. Bush from boycotting the Kyoto Protocol, it played an important role in galvanizing international cooperation to address climate change. The EU ratified the Kyoto Protocol in 2002. After the accession of Russia and Japan, the protocol came into force in 2005.

The fourth report of 2007 concluded on an even more confident and urgent note: "Most of the observed increase in global average temperatures since the mid-twentieth century is very likely due to the observed increase in anthropogenic GHG concentrations" (IPCC 2007a, 5). Nevertheless, the path from scientific expertise to political action was not straightforward, as shown by the abject failure of the Copenhagen Summit in December 2009. Despite hopeful declarations of intent made at the subsequent conferences held at Cancun in 2010 and Durban in 2011, the stalemate has not been overcome and in fact is unlikely to be overcome any time soon (Marsden 2011).

This policy failure is despite the fact that, since the 1980s, climate scientists have embraced post-normal science. The peer community has been extended to include political decision makers and other stakeholders. The IPCC's summaries for policymakers are approved line-by-line at plenary sessions of national delegates (Agrawala 1998; Saloranta 2001). Since climate change has become an issue of mass politics, media pundits and civil society have also become enmeshed in the extended peer community (Bäckstrand 2003).

Today, the extension of the peer community has reached an unprecedented level. An Internet-based analysis of the climate change community reveals the remarkable pluralism and vibrancy of this knowledge network and shows how closely it matches the model of post-normal science (figure 5.5).

Contrary to what one might expect, the IPCC is not the most important hub among the top fifty nodes of the climate change network. The IPCC ranks only tenth, after various blogs. Popular blogs such as scienceblogs.com and realclimate.org come in on top of the list and form

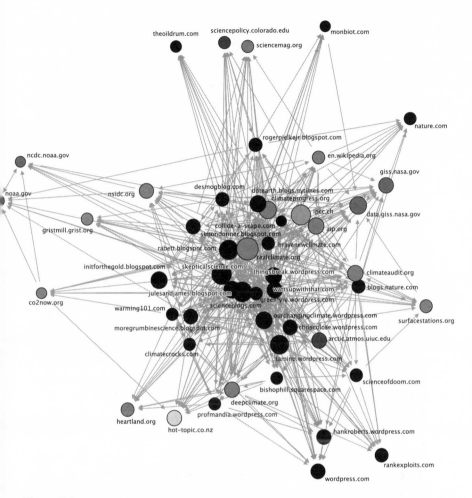

Figure 5.5
Internet-based cluster map of climate science. Analysis conducted on March 15, 2012, with issuecrawler.net. Starting points: ipcc.ch; realclimate.org; csmonitor .com/Environment/Global-Warming; skepticalscience.com; blogs.nature.com/cli matefeedback.

the largest group, occupying twenty-five among the top fifty nodes. While some of these climate blogs provide perspectives on the mainstream view, others are dedicated to debunking climate skepticism. As a side effect of the fierce polemic, the network contains as many as six climate-skeptical nodes. Only thirty-nine of the top fifty nodes in the network are largely devoted to propagating the mainstream view about climate change.

The magazines *Nature* and *Science* occupy an important position, but less so than the Wikipedia page about climate change. Only seven nodes are research institutes and data sources, while many more are websites run by committed climate scientists. Thus, at least in the blogosphere of the Internet, climate science in a narrow sense is eclipsed by the extended peer community.

Climate Skepticism
Climate science is a typical case of post-normal science caught in a stranglehold, or double whammy, between two different kinds of abnormal science: on the one hand radical alarmists criticizing the post-normal establishment; and on the other hand contrarian scientists alienated by the shift from what they would consider normal, "value-free" science.

The former group of radical alarmists holds that the climate consensus underestimates the gravity of the situation. To characterize their stance it is sufficient to cite book titles such as *The Revenge of Gaia: Why the Earth is Fighting Back* (Lovelock 2006), *Down to the Wire: Confronting Climate Collapse* (Orr 2009), *The Vanishing Face of Gaia: A Final Warning* (Lovelock 2009), and *Requiem for a Species: Why We Resist the Truth about Climate Change* (Hamilton 2010). A variety of apocalyptic movies, novels, and blogs amplifies the catastrophist streak (Urry 2011, 36–47).

This need not detain us much because alarmism about climate change is inherently similar to alarmism about peak oil. The only relevant difference is that, due to previous links and long-standing affinities, radical alarmists about climate change are better connected with the new post-normal mainstream than peak oilers, who are struggling in complete isolation from mainstream science.[19]

Radical alarmism aside, it will be more interesting to focus on the second category of contrarian scientists acting as "climate skeptics." Before doing so, however, it is fair to provide a clarification regarding terminology. Skepticism is rightly seen as a scientific virtue (Merton 1973), and climate skeptics give it a bad name. Nevertheless, I adopt this term here because it is common parlance and because I do not wish to discriminate against people propounding arguments in a debate. Climate skeptics openly talk about climate change, and in that sense they are not in denial. Therefore, I reserve the term "climate change deniers" for people who altogether avoid the issue (see chapter 6).

To understand the conversion of former adherents to normal science into contrarian climate skeptics, it is important to note that, even before the shift to post-normal science, climate science was far from apolitical.

To begin with, there were firm links between meteorologists and the military establishment in the first couple of decades after World War II. At the time military planners were keenly aware that accurate weather forecasting had decided important battles such as the landing in Normandy. What is more, some Cold Warriors nurtured the technological vision of obliterating the Soviet Union via climatic warfare. Finally, the early study of oceanic and atmospheric systems was intimately linked with the study of nuclear physics, especially nuclear fallout (Weart 2008, 20–22, 54–55, 133).

As a consequence, climate scientists in general and physicists in particular enjoyed privileged access to research funds and executive decision making. Even though normal climate science was never apolitical, as long as the political status quo persisted and the issues in question were not dragged into the limelight, the political links were largely invisible to the public and even to most scientists. That comfortable state of affairs was radically altered by the environmental and antiwar movements of the 1960s and 1970s, which challenged the presumption that science and technical progress are apolitical.

This led to a series of political and administrative shake-ups, as a result of which establishment scientists lost much of their privileged access to research funding and executive decision making. Some of them understandably resented their loss of status and became fierce opponents of post-normal science. This was particularly true about physical scientists who had previously been among the most privileged (Oreskes and Conway 2010). A prominent example of backlash from former establishment scientists was the trio of physicists—Frederick Seitz, Robert Jastrow, and William Nierenberg—who in 1984 founded the George Marshall Institute, which became a major player in the climate-skeptical movement (Lahsen 2008).

In the conservative establishment, the first reaction was a head-on clash with environmentalism that culminated with the Reagan administration. However, it soon became clear to Republican strategists and business interests that an openly anti-environmental stance was not appreciated by the median voter.

Therefore, conservative interests increasingly supported contrarian scientists who undermined the science invoked by environmentalists to justify progressive policy prescriptions. In particular during the Republican revolution waged by Newt Gingrich against the Clinton administration, this somewhat more subtle approach culminated in the skepticism about, and sometimes the outright rejection of, unwelcome scientific findings (Ehrlich and Ehrlich 1998).

While the climate consensus became increasingly accepted in other parts of the world, most notably in Western European countries such as Germany, climate skepticism further intensified in the United States, where contrarian scientists received strenuous support from conservative stakeholders in the business community and the political establishment (Grundmann 2007).

Ever since the inception of climate skepticism in the 1980s, US conservatives have opposed market intervention on ideological principle and followed what may be called the Republican *modus tollens*: "If the scientists are correct, it means certain actions should be taken; those actions should not be taken; therefore, the scientists are not correct" (Mazo 2011, 240).

During the George W. Bush administration, conservative think tanks and hard-line pundits persisted in their ideologically motivated opposition to mainstream climate science (McCright and Dunlap 2010). They did this despite the fact that business interests increasingly distanced themselves from climate skepticism. For example, the Global Climate Coalition, founded in 1989, was disbanded in 2002. Only a small minority of American businesses such as Koch Industries and, for a long time, Exxon, defied the trend and continued to support climate skepticism (Greenpeace 2010).

Despite the reduced level of industry support, there is now a well-oiled machine promoting organized climate skepticism. This has included the creation of front groups and fake grassroots organizations, called Astroturf groups, to deceptively spread the notion that climate skepticism has become a popular movement. All of this has been amply documented by other authors, so I can be brief here (see Dunlap and McCright 2011; Washington and Cook 2011; Oreskes and Conway 2010; Hoggan 2009; Monbiot 2006).

What is less well documented is that, after the end of the George W. Bush administration, organized climate skepticism has received unexpected reinforcement from two sides: the populist free marketers of the Tea Party, and climate-skeptical bloggers. Bloggers in particular are incredibly influential and have successfully spread climate skepticism from its traditional "homeland" in the United States to other Anglo-Saxon countries such as Canada, Australia, and the United Kingdom, as well as other parts of the world.

This climate-skeptical grassroots movement of "lay experts" and amateur bloggers attacking the consensus on global warming had its beginnings in the mid-2000s. While only a minority of these cyber-skeptics has formal credentials in climate science, the most prominent bloggers, such as Steve McIntyre and Anthony Watts, do have a relevant scientific background. Although one may assume that skeptical bloggers sometimes benefit from organized climate skepticism, for example by being invited as speakers by conservative think tanks or by attending their conferences, the considerable prominence of such bloggers does not depend on direct financial support (Zajko 2010, 2011).

Thus, climate-skeptical blogging constitutes an authentic reaction to post-normal science. In the heated atmosphere of the climate controversy of the late 2000s, this has prompted mainstream climate scientists and sympathetic amateurs to counter climate-skeptical bloggers with their own websites and blogs. The result is a cacophony of dueling blogs and websites. For example, Steve McIntyre's climate-skeptical site, climate-audit.org, is purposefully countered by an anti-skeptical website called realclimate.org.

A network analysis of the climate-skeptical community yields the following picture (figure 5.6). The top fifty nodes in the network are

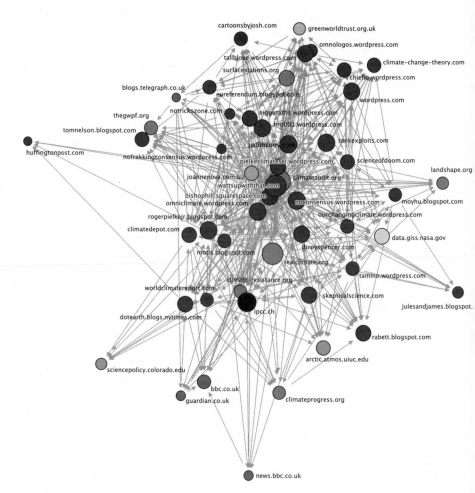

Figure 5.6
Internet-based cluster map of climate skepticism. Analysis conducted on March 21, 2012, with issuecrawler.net. Starting points: wattsupwiththat.com; climateaudit.org; thegwpf.org; judithcurry.com; bishophill.squarespace.com.

dominated by twenty-five climate-skeptical websites, twenty of which are blogs. The dueling approach between climate skeptics and those trying to debunk them explains why the twenty-five climate-skeptical websites are matched by fourteen climate-convinced websites, nine of which are blogs.

As we have seen (figure 5.5), this is mirrored by an analogous presence of climate-skeptical websites in the network cluster representing mainstream climate science. There definitely is engagement between post-normal climate science and its contrarian critics, although the style of the confrontation is mostly hostile. Even negative citations and outright polemics enmesh the climate-skeptical with the climate-convinced network, and vice versa.

The result is that climate skepticism looks far more authoritative than could be explained by the fact that about 2–3 percent of climate scientists subscribe to contrarian views (Anderegg et al. 2010). Contrarian scientists are typically less published and less cited than their peers, and yet the media award them disproportionate attention compared to the 97–98 percent of climate scientists subscribing to the consensus view of global warming. This is particularly true in Anglo-Saxon countries and especially in the United States, where climate change has become an object of enormous polarization and a political football in the acrimonious confrontation between conservatives promoting a free market ideology and liberals calling for state intervention (Painter 2011; McCright and Dunlap 2011; Hoffman 2011).

The Double Bind of Post-Normal Science

Insofar as the continued viability of industrial society is at stake, climate change and energy scarcity are clearly post-normal problems. Some observers have been tempted to conclude that such problems require an overtly political form of knowledge production, or post-normal science (Healy 1999; Ravetz 2004).

While this has significant appeal, the shift from normal to post-normal science is far from automatic. It has happened in the case of climate change, but not to the same extent in the case of energy. The different patterns of contestation are inherently interesting, but they also need to be explained. Moreover, it is an open question whether post-normal science is any better placed than normal science to deal with post-normal problems. For example, it seems hard to deny that normal energy science abjectly fails to confront the risk of energy scarcity. But has post-normal climate science brought us any closer to seriously addressing, let alone "solving," the problem of climate change?

Different Patterns of Contestation

There are two different patterns of contestation between establishment science and abnormal science. In the case of energy, there is open hostility toward, but little interaction with, the radical alarmists. In the case of climate change, interactions between establishment scientists and radical alarmists are more intense and amicable. However, the same does not hold for contrarian scientists. Here, the relationship is deeply acrimonious.

In the case of energy, a normal science heavily dominated by economists continues to hold that there is no fundamental problem, despite the moaning of alarmist "cranks" from the peak oil community. In the case of climate change, by contrast, various disciplines have reassembled themselves into a post-normal science open to like-minded stakeholders depicting global warming as a serious threat to humankind. As a consequence, in the case of climate change it is the contrarian scientists objecting to the "negative" views of post-normal science who are widely seen as the "cranks."

Despite all the bad blood, there is an intense level of interaction between post-normal climate science and climate skepticism. This is the result of the dueling approach: the proponents of climate science and their opponents constantly refer to each other. As we have seen, the top fifty nodes in the climate change network (figure 5.5) include six climate-skeptical sites, while the top fifty nodes in the climate-skeptical network (figure 5.6) include sixteen websites defending the global warming hypothesis. Altogether twenty nodes are common to both networks. In both climate networks the density of linkages is tremendous, as indicated by the gray lines connecting the nodes.

Now compare this to the energy networks (figures 5.3 and 5.4), where the density of intra-network interactions is much lower and where there are only four nodes common to both the energy supply and peak oil networks. Apparently, post-normal science leads to greater vibrancy and engagement than the artificial separation between normal and abnormal science.

Another finding confirming our models (figures 5.1 and 5.2) is that, in the case of energy scarcity, there are virtually no vested interests supporting the alarmists. Because post-normal science implies more widespread political polarization, this is different in the case of climate change. Climate scientists are supported by progressive interests, whereas climate skeptics can rely on moral and financial support from conservative groups.

By the way, there are only a few links connecting the climate and energy networks. The climate science network includes *The Oil Drum* as number 42, and the peak oil network includes the IPCC as number 36 of the top 50 nodes of the network. Some of the bloggers and activists in the climate science network, such as George Monbiot, are also concerned about energy scarcity; and some peak oil activists, such as the adherents of the Transition Town movement, are also concerned about climate change. Apart from these tenuous linkages, however, the two fields of knowledge lead a surprisingly separate existence.

Explaining the Difference

So why has post-normal science prevailed in the case of climate change but not in the energy case? This has to do with the different situation experienced by the scientists. The study of climate change has become a post-normal science because the academic disciplines concerned were attracted by a new research agenda emerging with the environmental movements of the 1960s and 1970s. In the energy case, the absence of a sufficiently attractive alternative research program, as well the dominance of economists and engineers, prevented energy science from becoming post-normal.

Climate science literally offered a golden opportunity for scientists to transcend a variety of mother disciplines, from geophysics to geochemistry and from meteorology to oceanography, in order to study a fascinating phenomenon. The integration of non-scientific stakeholders, from policy makers to advocacy groups, enabled access to the necessary funds.

The study of energy did not offer the same set of opportunities. Energy can be studied by individual scholars as a fascinating trans-disciplinary field (for an outstanding example, see Smil 2008a), but unlike climate change it does not offer a well-funded platform to integrate various disciplines into one multidisciplinary research agenda. Therefore, it is understandable that energy experts continued to rely on existing business and policy links. Moreover, they typically have a background in economics. Economics is far more conservative and far less amenable to environmentalism than most of the disciplines that constitute climate science. As a consequence, energy experts have largely left the issue of energy scarcity to mavericks at the fringes or outside mainstream science.

Another important part of the explanation is the fact that atmospheric CO_2 concentrations are continuously rising, whereas energy prices are highly volatile and were on a twenty-year low from mid-1980s to the mid-2000s. In addition to that, more and more hard data on climate change have accumulated whereas data on energy reserves remain shaky to say the least. As a consequence, for empirical reasons the risk of climate change has been somewhat more amenable to alarmism than the risk of energy scarcity.

Thus, the realm of climate science has been taken over by Cassandra while energy remains in the domain of Dr. Pangloss, the proverbial arch-optimist. Tragically, the policy outcome is virtually the same in either case. While normal energy science keeps the post-normal challenge of energy scarcity from the agenda, post-normal climate science is unable to translate the so-called climate consensus into political action because there are subtle mechanisms by which its authority within society at large is undermined.

Post-Normal Science as a Trap

Despite their embrace of post-normal science, or perhaps because of it, climate scientists find themselves torn between the reflexive culture of post-normal science and the continued validity of standard scientific values. Leading climate scientists such as Stephen Schneider and James Hansen noted early on the difficult balance between loyalty to the scientific method and the goal of policy relevance in the public sphere. Schneider has called this dilemma the "double ethical bind" of trying to balance practical effectiveness with intellectual honesty (Schneider 1988; Pool 1990; Russill 2010).

In fact, post-normal science imposes upon climate scientists an arduous balancing act in which they must invoke scientific objectivity to maintain their expert authority, while at the same time engaging in "stealth issue advocacy" to achieve policy impact (Pielke Jr. 2007). Thus, they are torn between the requirements of political interventions in the style of post-normal science, and the need to keep a posture of scientific objectivity. The reason underlying their dilemma is that, on the one hand, extending the peer community has politicized scientific discourse and galvanized part of the public for action, while on the other hand key sectors of the public do not forgive any dilution of scientific rigor—especially when it comes to "inconvenient truths."

Even among mainstream scientists, post-normal science amounts to a stigma rather than a neutral label, as for many the term suggests lower quality and an unclear separation between subjective opinion and objective analysis. Therefore even those tolerant to, or accepting the inevitability of, post-normal science, often prefer not to be associated with it themselves.[20]

The IPCC itself has adopted three strategies to maintain the veneer of adherence to standard scientific norms and thus hedge against the loss of expert authority that may result from engaging in post-normal science. First and foremost, the IPCC sticks to a painstaking procedure of peer review. Although that procedure is somewhat different from standard peer review in academic journals, it has protected the IPCC's scientific credibility.

Second, the IPCC attributes percentage values to the levels of confidence by which its findings are stated to be true (Saloranta 2001; Swart et al. 2009). For example, the expression "very likely" corresponds to a confidence level of at least 90 percent (IPCC 2007a, 27). The practice is somewhat reminiscent of statistical techniques and thereby conveys scientific objectivity.

Third, the IPCC is ostensibly committed to the fact–value distinction. Science conventionally prides itself on objective knowledge, which is assumed to have validity regardless of subjective values. To protect their scientific reputation, most climate experts insist on their value neutrality and are reluctant to translate their findings into policy prescriptions.

They claim to be able to reconstruct and, to some extent, predict climate change; but they hasten to submit that they are unable to answer the question of how much climate change is tolerable, or what policies ought to be adopted.

Such posturing may have temporarily shielded climate scientists from the allegation of stealth issue advocacy, but not for long. After all, the whole point of post-normal science is to be policy relevant. No matter how much climate scientists insist on their value neutrality, political decision makers need specific prescriptions to legitimize their policy choices. In a somewhat oblique way, climate scientists have obliged by offering focal points. For example, discussions became much simpler after consensus had been reached on 2°C as the highest tolerable increase of global temperature. Climate scientists did not actively promote this somewhat arbitrary target, but they put a good face on the matter when it was attributed to their expertise.

Needless to say, even post-normal science cannot tackle a civilizational problem of epic proportions such as climate change. But post-normal science may carry the seeds of its own destruction because authority continues to be ascribed to conventional, specialist science. While extending the peer community and thus diluting rigorous standards may or may not lead to more democratic processes, it does undermine the authority and tarnish the reputation of climate science. Populist climate skeptics relish in exposing the contamination of science with political motives, and particularly the political nature of the alliance between scientific experts and promoters of vigorous action to control climate change (McGaurr 2009).

The case of climate change thus shows not only the potentialities of post-normal science but also its limitations. On the positive side, when science is politicized and becomes post-normal, public awareness is increased. To the optimist, this indicates that modernity has become reflexive about the unintended consequences of its own success (Beck 1992, 1999, 2009). On the negative side, post-normal science unleashes a reaction of contrarian scientists supported by social, political, and economic forces opposed to its policy prescriptions. The prescriptions derived from post-normal science are easily thwarted by this "anti-reflexive" impulse (McCright and Dunlap 2010).[21]

At the end of the day the eagerness of climate scientists to be policy relevant has created a serious challenge to their reputation. A few isolated cases of compromised integrity, such as the leakage of emails from the University of East Anglia and a couple of overstatements in the 2007 IPCC Assessment Report, like the rapid melting of Himalayan glaciers, have been sufficient to thoroughly discredit post-normal climate science.

The proponents of post-normal science often take an a priori normative stance and claim that the democratization of science is leading to more adequate policies to address post-normal issues (Bäckstrand 2003). But is this really true? To what extent has post-normal science actually advanced the causes of climate change mitigation and adaptation? From a policy perspective, there is little to suggest that the turn from normal to post-normal science has actually helped to address climate change.

In principle, both climate change and energy scarcity are post-normal issues where "facts are uncertain, values in dispute, stakes high and decisions urgent" (Ravetz 2004, 349). Despite this similarity, post-normal science has come to dominate only climate change and not energy. At first glance this makes an enormous difference because, while energy science has kept the post-normal issue of energy scarcity off the agenda, climate science has successfully placed the equally post-normal issue of climate change on the public agenda.

Unfortunately, a closer look reveals that post-normal science does not translate into action when the issues at stake are too existential. This is not to deny that the IPCC has been uniquely successful. But key players prefer to ignore its findings or simply deny that climate change is taking place. Meanwhile, the inherent ambiguities of post-normal science have plunged climate science into a deep legitimacy crisis. Energy science has been spared a similar crisis, but as a result the risk of energy scarcity has for many years been withheld from official scrutiny.

Bottom Line

Considering the equally dismal outcome, scientists are caught in a double bind. On the one hand, there is an imperative to engage in post-normal science in order to promote awareness of civilizational problems. A failure to do so, as in the case of energy scarcity, is irresponsible given the momentous consequences of business as usual. On the other hand, when scientists engage in post-normal science as in the case of climate change, the awareness raised is only temporary. Over time, engaging in post-normal science undermines the fiction of scientific objectivity, which in turn erodes the authority and thus the ability of scientists to influence policy choices. Thus, scientists are damned if they do and damned if they don't engage in post-normal science.

6

The Moral Economy of Inaction

What prevents us, together and as moral individuals, from confronting existential problems such as climate change and energy scarcity? On the face of it, an effective response is hindered not just by the inadequacy of existing knowledge regimes but also by a confluence of behavioral and cognitive dispositions. Despite the inescapability of the impasse and the efforts of countless well-intentioned people and groups, there is a full-fledged moral economy of inaction to ensure that our response to climate change and energy scarcity falls short of what is required.

The moral economy of inaction consists of three key elements. First, people tend to greatly undervalue future events and distant strangers. The more remote somebody or something is from us, the less we care. We may call this ethical discounting, and it comes in two different forms, temporal and spatial. Second, the pursuit of particular interests often thwarts collectively desirable outcomes. We may all agree that something needs to be done, and yet not a single one of us may be ready to do it. This amounts to collective action problems, which occur when the pursuit of particular interests impedes collectively desirable outcomes. Third, people often simply pretend that their problems do not exist. This is also called denial, and it can be defined as the habit of treating real problems as if they were non-issues.

Denial in particular is an underappreciated but enormously important part of the moral economy of inaction. It has a rational core because it minimizes pain, but it often leads to tragic outcomes.[1] The nature and gravity of these consequences depends on whether a problem is tractable or intractable, and whether it is permanent or escalating. In general the denial of intractable problems is less harmful than the denial of tractable problems, and the denial of permanent problems is less harmful than the denial of escalating problems. Unfortunately, climate change and energy scarcity appear to be escalating problems.

Because denial can have extremely harmful consequences, I discuss the possibility of social intervention. What, if anything, can non-denialists do to liberate people from damaging forms of denial? I demonstrate that denial can sometimes be transcended by means of rational persuasion, but other modes of social intervention are less likely to succeed.

The situation is complicated by the fact that denial is interconnected with ethical discounting and collective action problems to form an integrated moral economy of inaction. I conclude by placing this moral economy of inaction in the broader context of humankind's common evolutionary heritage.

Ethical Discounting

As observed by David Hume in his *Treatise of Human Nature* (1739), the more distant something is in time and space, the less we care about it.[2] This implies that when our behavior is unsustainable in the long run, and/or when it is damaging to distant others, we have a tendency to discount the effects and focus on the here and now. This tendency may be called ethical discounting, and it comes in two distinct forms: temporal and spatial. Temporal discounting is a function of the time period between present decision making and future consequences, whereas spatial discounting is a function of the geographical or emotional distance between the people acting and the people affected.

Temporal Discounting
John Maynard Keynes famously cautioned that the long run is "a misleading guide to current affairs. In the long run we are all dead" (Keynes 1923, 80). The argument is that, no matter their inherent merits, long-term considerations are futile because our life happens now, and not in some distant future. We should therefore not unnecessarily worry about the future and instead put our focus on the events of the day. Whether or not this is morally desirable, it accurately describes the routine behavior of citizens, economic stakeholders, and even politicians who are notoriously concerned about (re)election.

Climate change and energy scarcity are cases in point. The effects of this year's CO_2 emissions will be felt many years from now, and the worst effects accrue to future generations. Similarly, no matter how prodigally we deplete resources now the worst effects of scarcity will occur in a distant future.

Economists have developed a fancy way of expressing the hidden rationality of preferring instant gratification over long-term benefits: *the rate of time preference*. For the sake of argument, let us assume you know that the roof of your house is not going to withstand next winter. How much would you pay *now* to fix it and thus prevent the expense of $10,000 *next year*? If we assume that the interest rate is 5 percent, then it would not be economically rational for you to pay more than $9,524. Otherwise, you could take the money to the bank and have more than the necessary $10,000 a year from now.[3]

In the hypothetical case just mentioned, you would be discounting the future by a rate of 5 percent per year. Now let us assume you are anticipating a loss of $10,000 forty years from now. Like in the case of the shaky roof, let us assume that your discount rate is 5 percent. If you do the math properly, you will find that you should not pay more than $1,420 now to ensure yourself against that future loss. This is less than 15 percent of the expected damage.[4]

If we translate this to societal problems, discounting is influenced not so much by interest rates but rather by the expectation of economic growth. Insofar as people will be wealthier in the future, repairing damage will become easier for them. Therefore, if we assume that the next generation will be twice as wealthy, we should fix a problem now only if doing so is less than half as expensive as it will be for the next generation. Otherwise, it makes more sense for the future to take care of the problems we are leaving behind.

It follows that we can calculate from our expectations of economic growth the rate by which we are entitled to discount future cost. Let us call this the *growth-based rate of time preference* because it is a measure of how much we prefer the present over the future based on anticipated growth of wealth or the standard of living. Apart from this growth-based time preference we may also discount the future for other, more egoistic reasons. Economists call this the *pure rate of time preference*. While economists agree that the growth-based time preference is on solid ground, the moral basis for the pure time preference is contested because there is no principled reason from a utilitarian standpoint why the present should carry more weight only because it happens to be the time we are living in.

According to the canonical formula associated with the mathematician Frank Ramsey (1928), the discount rate (r) is the sum of the pure time preference (δ) and the growth-based time preference, with the latter defined as the product of economic growth (g) and a parameter (η).[5] Thus, $r = \delta + \eta \cdot g$.

This way of thinking has made it possible to reduce the problem of climate change to a technical debate. For example, Nordhaus and Boyer (1999) assume a high value for pure time preference, thus exonerating decision makers from action to mitigate climate change. Others set the pure time preference close to zero. This has led Nicholas Stern (2007, 663) to recommend action to mitigate climate change. Others in turn have criticized this (Quiggin 2008). The debate is unlikely to end any time soon. It ultimately boils down to this. There are sound moral reasons why the pure time preference should be low, but there are equally sound empirical reasons why it actually is rather high.

There are also problems with the growth-based rate of time preference. Most economists simply assume business as usual, including further growth (Dasgupta 2008, 149). But we may easily imagine a serious and long-term economic contraction, maybe due to catastrophic climate change. This would lead to a negative shrinkage-based rate of time preference, turning the logic of temporal discounting on its head. It would suggest that we must address climate change now because future generations will be poorer and therefore less able to do it. While that sounds like a compelling conclusion from a logical viewpoint, only few economists are willing to contemplate it.[6]

Another problem is deep structural uncertainty. The links between the anthropogenic causes of climate change and systemic effects on human systems are so complex that after thirty years of modeling the consequences are still very hard to predict. Similarly, fuel depletion models are obfuscated by unreliable data and depends on hazardous assumptions about reserve growth and technological progress. The disruptive effects of a global energy crunch on the economy are almost impossible to estimate. Given such fundamental systemic uncertainty, there is an understandable tendency in many quarters to dismiss the risks of catastrophic climate change and disruptive energy scarcity as unfathomable "low probability high impact" events.

One economist, Martin Weitzman, has undertaken a heroic attempt to factor the unfathomable deep risks and structural uncertainties of catastrophic climate change into conventional economic modeling (2009, 2011). He shows that the potential effects of catastrophic climate change make it very challenging to apply standard cost–benefit analysis. While Weitzman is still trying to reform economic modeling to incorporate deep risk and structural uncertainty, it is perhaps fair to draw an even more radical conclusion: existential civilizational predicaments such as climate change, or energy scarcity for that matter, are simply not amenable to economic modeling.

This does not mean that temporal discounting is not taking place. Instead, it simply follows a less rational but tragically all-too-human form. When deeply engrained habits are at stake, humans tend to discount the future much more radically than any utilitarian model would predict.

For example, a smoker prefers the risk of a miserable death twenty years from now to abandoning her habit. This is partially explained by the behavioral pattern of procrastination (Akerlof 1991). The smoker decides today that she will quit tomorrow, but when tomorrow comes she decides that she will quit the day after, and so on. She may of course regret this eighteen years from now.[7]

Arguably, prodigal use of energy and the concomitant CO_2 emissions are deeply engrained habits at the civilizational level that are roughly analogous to the personal vice of smoking. It is easy to agree that civilization should kick these habits, but taking the necessary immediate steps does not come easily.

The longer the time frame, the greater the uncertainties, and the deeper the habit, the more future damage is accepted for the sake of instant gratification, and the less likely preventive action is undertaken. Such radical discounting of the future may be irresponsible, but it is incontrovertibly human.

Spatial Discounting

The other form of ethical discounting is spatial discounting. Sometimes this means literally discounting utility based on geographical distance

(Perrings and Hannon 2001). More commonly, however, people discount on the basis of emotional rather than kilometrical distance. The main discounting criterion is perceived personal proximity or remoteness, which in turn may be determined by discriminations made on the basis of class, race, gender, and so on.

Most of us are willing to support close relatives in need, but our willingness to help distant strangers is more limited. As citizens we grudgingly accept that a considerable part of our income is taxed away for social redistribution, but most of us would not accept more than a marginal amount of our taxes to be spent on development aid. This is not to deny that some of us are ready to give when the misery of distant strangers is brought to our attention through heartbreaking human interest stories. But the reason, then, is precisely that our emotional remoteness from distant strangers is temporarily reduced.

Once again, climate change and energy scarcity are cases in point. While people in industrialized countries have the highest carbon footprint, they may be tempted to assume that the worst effects of climate change happen in remote developing countries. Similarly, while the high per-capita resource consumption of people in rich countries may lead to exorbitant prices for fuel and other commodities, consumers in poor countries are hit hardest by these prices while people in rich countries are able to afford them.

All of this is part of the human condition. We are more than animals but still very brute. We can feel our own joy and pain, and perhaps some of the joy and pain of the people close to us. In times of globalization our actions affect distant strangers, but we are separated from them by impersonal transactions. We can fantasize about universal love and cosmic fellowship across space, time, and species. But apart from rare moments of enthusiasm we cannot really feel the connection. Only God can do that. For ordinary mortals like you and me, even loving our neighbor is a big challenge.

Figure 6.1 summarizes the effect of ethical discounting, both temporal and spatial, on the likelihood of people caring about an issue.

Collective Action Problems

When individual actors maximize gain, profit tends to be private but negative externalities are shared. When free to do so, rational actors

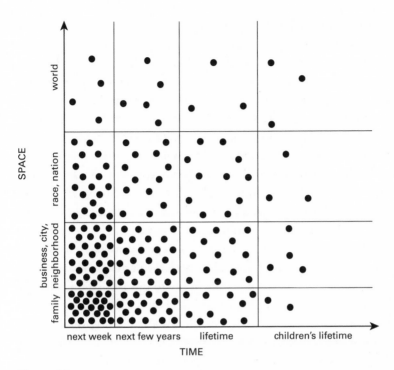

Figure 6.1
Ethical discounting. "Although the perspectives of the world's people vary in space and in time, every human concern falls somewhere on the space-time graph. The majority of the world's people are concerned with matters that affect only family or friends over a short period of time. Others look farther ahead in time or over a larger area—a city or a nation. Only a very few people have a global perspective that extends far into the future." *Source:* Meadows et al. 1972, 19.

maximize private utility regardless of the social effects. As a consequence, public goods are in jeopardy. The classical example is the so-called tragedy of the commons, where communal grazing land is ruined by the scramble of individual herders for a declining land base (Hardin 1968).

Climate change mitigation and sustainable energy use can be seen as public goods, or global commons (Stern 2007). If they are achieved, everybody benefits and nobody is left behind. Unfortunately, however, collective action in favor of global public goods does not come easily (Levin 2010). Even if there were consensus that vigorous action is needed to prevent their further degradation, there would be serious collective action problems.[8]

At various levels of agency, collective action is made difficult by social traps such as the free rider problem and the prisoner's dilemma. It is enormously tempting from the viewpoint of individual countries or citizens to continue business as usual while hoping that others will invest in climate change mitigation—that is, to free-ride. By the same token, as users of scarce energy resources we are in a prisoner's dilemma: it doesn't make a difference for my own future access to these resources if I conserve energy, unless all others conserve energy too; but because I know that they are unlikely to do so, it would be foolish for me to conserve energy. Similarly, it is perfectly rational for a national government to call for a global climate deal while hammering out a growth package for the domestic economy. Because individually all actors are tempted to free-ride and succumb to the prisoner's dilemma, at the collective level cooperation is unlikely to occur—and fragile when it does.

All of these are examples of collective action problems where a publicly desirable outcome cannot be reached because the participants are driven by the self-interested pursuit of private utility to thwart that outcome. In theory, there are three possible ways to overcome such problems: hierarchical solutions, market-based solutions, and societal self-governance.

In the current climate, the feasibility and effectiveness of hierarchical solutions is contested. This is not to deny that many observers call for binding international agreements, such as an international convention to limit carbon emissions. The distinguished British sociologist Anthony Giddens (2009) calls for a more political approach to climate change. However, given the absence of a robust international consensus, most observers are skeptical about the prospects for top-down governmental solutions.

It is more in vogue to hope for market-based solutions. Some set their hope in voluntary schemes such as corporate social responsibility or consumers making donations to offset their carbon footprint (Newell and Paterson 2010). Others call for home owners to voluntarily switch from cheap electricity based on fossil fuels to more expensive electricity produced from renewable energy. While this is appealing from a free-market perspective, it does not even begin to address the problems precisely because collective action problems are at play. Even in rich countries, carbon offsetting and corporate social responsibility work only in niche markets, and there are only few "ethical consumers" willing to pay a premium on their power bills when there is no guarantee that enough of their fellow consumers will do the same.[9]

Therefore, market-based solutions must be backed by hierarchical intervention. This is exactly what happens with emissions trading, carbon taxes, and feed-in tariffs. All of these are the result of top-down political regulation. Once again, however, there is a collective action problem: costly climate regulation may lead to carbon leakage whereby carbon intensive production simply moves to jurisdictions where there is no such costly regulation. Therefore, to work on a global scale, these instruments would require a binding international agreement. As we have seen, however, such an agreement is not forthcoming.

Much has been made of the fact that traditional societies and local communities sometimes develop institutional mechanisms to overcome the tragedy of the commons (Ostrom 1990; Poteete, Janssen, and Ostrom 2010). While this is undoubtedly true, unfortunately it is inapplicable to global problems such as climate change and energy scarcity. Bottom-up solutions to collective action problems require a very demanding social fabric, which is sometimes present in traditional societies and local communities but clearly absent at the global level of industrial civilization as a whole.

Another serious problem related to climate change and energy scarcity is the lack of credible and effective global leadership. With a credible and effective global leader, it is sometimes possible to establish institutional devices to overcome collective action problems. For example, international free trade was reinstated after World War II under the hegemonic leadership of the United States (Kindleberger 1973; Gilpin 1987). Once established, international regimes can then develop a life of their own and stabilize a situation even in the absence of global leadership (Keohane 1984).

In the cases of oil depletion and climate change, the most desirable candidates for global leadership are the United States and China because they are the most important consumers and the most significant polluters.[10] Unfortunately, they both appear to be part of the problem rather than part of the solution. Absent their leadership, it is hardly surprising that international organizations such as the IEA and the IPCC have not made a big difference.

The Twisted Rationality of Denial

Another cornerstone in the moral economy of inaction is denial, or the habit of treating a real problem as if it were a non-issue.[11] A real problem is one that makes us suffer regardless of whether or not we acknowledge it. Denying such a problem seems perverse from a moral and ethical viewpoint, and it may easily have pernicious consequences. And yet, as we shall see, denial has its own twisted rationality as a strategy of pain avoidance and harm minimization.

Denial is a ubiquitous social and psychological phenomenon (Goleman 1985; Cohen 2001; Zerubavel 2006). Any kind of problem, from personal trauma to planetary challenges such as climate change and energy scarcity, can be obfuscated by denial. Denial can be seen when an individual disavows a terminal illness (Weisman 1972). It can also be studied in social constellations, from the denial of marital infidelity to the denial of race discrimination. Some cases are deeply political, for example when a nation state such as Turkey denies genocide (Alayarian 2008). A particularly worrisome manifestation is found at the global level, where most of humankind is in denial of the fact that infinite growth on a finite planet is impossible (Orr and Ehrenfeld 1995; Meadows, Randers, and Meadows 2004).

There are good reasons to see denial in a negative light. Disavowing one's problems is both cowardly and dishonest, and it often gets in the way of finding and adopting effective solutions. The consequences can be disastrous. As one author notes, "Denialism has already killed. AIDS denial has killed an estimated 330,000 South Africans. Tobacco denial delayed action to prevent smoking-related deaths. Vaccine denial has given a new lease of life to killer diseases like measles and polio. Meanwhile, climate change denial delays action to prevent warming" (MacKenzie 2010, 41).

Despite such perverse outcomes, denial is not irrational. When in denial, people follow what they feel to be in their best interest by minimizing real or perceived harm, thereby maximizing subjective and/or intersubjective wellbeing. Acknowledging a problem may lead to considerable psychological and social cost: negative emotions such as fear, shame, and helplessness; cognitive dissonance; loss of identity, or loss of friends; embarrassment; and social conflict about the attribution of blame and responsibility. Many people have a predisposition to minimize such psychosocial cost by establishing regimes of denial, rather than relentlessly facing up to their problems. Such behavior may be short-sighted and morally dubious, but it is by no means irrational.

At its core, denial is best understood as a psychosocial coping mechanism. Precisely because humans are psychic and social beings, problems are not simply "out there": acknowledging them carries psychosocial costs. Facing the alternative of avowing or denying a problem, people are driven to choose the easier option. Thus, denial is based on a self-interested rationale of pain avoidance and harm minimization.

When denial is perceived as less costly than avowal, there are good utilitarian reasons to deny the existence of a problem. This does not alter the fact that, depending on the characteristics of the issue, denial can have deeply ironic effects. For example, in treating a real problem as a non-issue, people often get themselves into a tussle of short-term versus long-term considerations. From a short-term viewpoint, denial is preferable when it is less costly than avowal *at any given moment*. From a long-term viewpoint, it is preferable when it lowers the cumulative cost incurred *over the whole duration of the problem*.

Overall, it seems fair to say that denial has a rational core but often leads to problematic side effects. The nature and gravity of these negative side effects depends on whether a problem is tractable or intractable, and whether it is permanent or escalating. The denial of intractable problems is less harmful than the denial of tractable problems, and the denial of permanent problems is less harmful than the denial of escalating problems. This is captured by the following climax from harmless escapist to genuinely fateful denial.

A. Escapist denial (of intractable and permanent problems) is beneficial both from a short-term and a long-term perspective. If there is nothing we can do about such a problem, it may indeed be better to forget.

B. Fatalist denial (of intractable and escalating problems) is beneficial when seen from a short-term perspective, but in the long run it leaves people in a greater mess and prevents them from cutting their losses.

C. Fateful denial (of tractable and escalating problems) is even worse because, apart from leaving people in a greater mess and preventing them from cutting their losses, it also prevents possible solutions.

It is easy to see that climate change and energy scarcity are escalating problems, so the category of escapist denial can be discarded (although it is a useful backdrop for understanding fatalist and fateful denial). Depending on whether or not we see them as tractable, they either fall in the category of fatalist or of fateful denial. While there is room for reasonable disagreement on this point, my own view is that they are largely intractable. Precisely because, deep down, people know that climate change and energy scarcity are existential predicaments that cannot be solved within the existing paradigm, they have a tendency to indulge in denial and deceive themselves that, for whatever reason, it is expedient to treat these problems as non-issues.

Escapist Denial

Escapist denial is the denial of a permanent intractable problem. It is the most benign form of denial, insofar as it is fully rational both in the short run and in the long term. Although the denial of climate change and energy scarcity is clearly not of this harmless kind, escapist denial is a useful starting point for our discussion as it exposes the logic of denial in simple terms.

Folk wisdom has it that, when a problem is intractable, denial is a reasonable strategy to deal with it. The following statement represents a popular philosophy of dealing with intractable problems: "If we can do something to solve a problem, we will perhaps do it. If there is nothing we can do, we should try to forget the problem." And indeed: When an intractable problem is permanent, then denial is entirely rational as a way for people to evade the discomfort of acknowledging the predicament in which they are caught.

Let us take a simple example: Jane is being teased at school for her ugly nose, which deeply humiliates her. One fine day she meets Jim, who never even mentions her nose. Instead, he makes compliments about her angelic face. Jane knows it is only sweet talk, but it makes her feel so good. She marries Jim, who knows full well how much she loves his compliments. He keeps oozing charm on his spouse, and this way their marital life turns out far better than it would otherwise be.

This is illustrated by figure 6.2. In the model, a problem appears at t_0. The problem causes direct cost ($cost_{problem}$). However, the total cost of the situation ($cost_{avowal}$) is higher due to the psychological and social cost resulting from avowal. A regime of denial is established at t_1, when disavowing the problem becomes possible. While the direct cost of the problem remains unaltered, the total cost of the situation is somewhat reduced ($cost_{denial}$).[12] The darker-shaded area indicates the cumulative cost

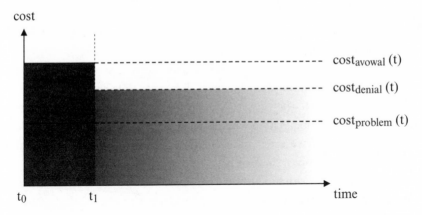

cost

Figure 6.2
Escapist denial

incurred until t_1, while the lighter-shaded area indicates the cumulative cost incurred thereafter. Fading shades and dashed lines indicate that deniers conceal the situation from each other and from themselves.

The model shows that when denying a permanent and intractable problem is less costly than acknowledging it, escapist denial has the desirable effect of minimizing cost. After t_1, the cost incurred at any given point in time is lower than before. As a consequence, escapist denial is desirable from a short-term viewpoint. It is also desirable from a long-term perspective because the cost incurred over the whole duration of the problem is equally lowered.

Long-term cost for avowers: $\int_{t_0}^{t} cost_{avowal}(t)$

Long-term cost for deniers: $\int_{t_0}^{t_1} cost_{avowal}(t) + \int_{t_1}^{t} cost_{denial}(t)$

Denial is better than avowal when: $\int_{t_1}^{t} cost_{denial}(t) < \int_{t_1}^{t} cost_{avowal}(t)$

As the fading shades and dashed lines in the model suggest, denial has a numbing effect. This becomes clear if denial is contemplated not only as a cost-saving device but also as a mental habit and social institution. Experimental psychology suggests that, once a habit has been established, it usually becomes automated unless and until there is a major disruption in the context of action (Ouellette and Wood 1998). Similarly, a key insight of new institutional economics is that, once a cost-saving social institution has been established, it is rarely questioned (David 1985; North 1990, 1991).

All of this is particularly true about denial, which is a mental rather than a behavioral habit and thus even less accessible to control. When people are in denial, they cover their problems with a mantle of silence. Even the fact of being in denial becomes an object of denial. For example, if Jim and Jane have erected a taboo around her ugly nose, then they can also not acknowledge that there is such a taboo. As a consequence, denial obfuscates the true situation not only from outsiders but from the deniers themselves (Goleman 1985).

Fatalist Denial

Climate change and energy scarcity are escalating problems, so the model of escapist denial does not apply. Escalating problems are problems that have an inherent tendency to spiral out of control. Since there is a tolerance limit to how much people can take or survive, any strategy to cope with such problems is destined to break down at some point. When an intractable problem escalates, breakdown must happen regardless of whether or not people acknowledge the problem, because at some point the tolerance limit will be exceeded. Such situations are the perfect breeding ground for fatalist denial.

Fatalism is a defiant attitude in the face of escalating problems that cannot be solved. While people cannot control the escalation itself, they usually have an incentive to avoid the social and psychological cost of acknowledging their predicament by pretending that the escalation is not happening. In addition, there is another reason why people engage in fatalist denial: their tolerance limit is reached a bit later than if the situation were openly acknowledged.

To illustrate this, let us imagine a couple that has a problem of escalating debt. They both agree that their prodigal lifestyle is not negotiable. Consequently, there is no solution to their problem. On top of the cost of servicing their debt, they incur the social and psychological cost of acknowledging the fact that they are living beyond their means. Since this is highly unpleasant, at some point (t_1) they stop talking about their mounting debt. They feel much better. Their debt keeps rising, but instead of lamenting their problem they have now swept it under the carpet.

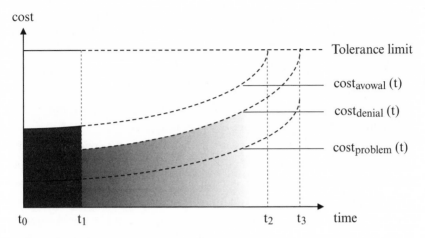

Figure 6.3
Fatalist denial

Initially, this seems to work very well indeed. But over time there is more and more fallout from their skyrocketing debt, and at some point (t_3) the entire situation is doomed to break down.

Figure 6.3 expresses the pattern in formal terms. In this model there is an escalating problem, with breakdown bound to occur when the total cost resulting from the situation hits the tolerance limit. As usual, denial is expedient insofar as it reduces social and psychological cost. If no regime of denial is established, the tolerance limit will be reached at t_2. If a regime of denial is established at t_1, then the social and psychological cost resulting from the situation is somewhat reduced. As a consequence, the situation can be protracted a bit longer until the tolerance limit is finally reached at t_3.

Thus, fatalist denial has two advantages: it reduces cost and it buys time. It is possible to express these advantages in mathematical terms.

Long-term cost for avowers: $\displaystyle\int_{t_0}^{t_2} cost_{avowal}(t)$

Long-term cost for deniers: $\displaystyle\int_{t_0}^{t_1} cost_{avowal}(t) + \int_{t_1}^{t_3} cost_{denial}(t)$

Denial is better than avowal when: $\displaystyle\int_{t_1}^{t_3} cost_{denial}(t) < \int_{t_1}^{t_2} cost_{avowal}(t)$

Alas, despite the benefit of reducing discomfort and buying time, there are three reasons why fatalist denial may lead to long-term disadvantages. First, it is far from clear that fatalist denial leads to lower cost incurred over the entire duration of the problem. In fact, the opposite may be true because fatalist denial extends the duration of the problem. It may easily happen that the total cost incurred over the duration of the problem is increased by the fact of being in denial.[13]

Second, fatalist denial leaves people with a greater mess. When the situation breaks down, the cost of avowal/denial disappears but the fallout from the problem is even higher. Since the problem keeps growing from t_2 to t_3, deniers end up worse than avowers.[14] For example, the consequences for the prodigal couple in our example are bad enough if they make their oath of disclosure at t_2. But if they stay in denial until t_3 when the bailiff finally knocks at their door, they will end up with an even greater burden of debt.

Third, even escalating intractable problems can be mitigated. But when people are in denial, they conceal from themselves essential information such as their own cost functions and tolerance limit. Like other deniers, fatalists hide from themselves even the very fact that they are in denial. Thus, they cannot anticipate the moment when the whole situation will come to a head. If the couple in our example ignore their bank statements and payment summonses, they will become fully aware of their predicament only when the bailiff comes to take their belongings away. This is too bad because, had they not been in denial, they could have second-guessed the moment when servicing their debt was going to become impossible. As a consequence, they would have been able to bunker cash and hide some of their belongings before the arrival of the bailiff. Having removed the problem from their consciousness, however, they are bound to miss any opportunities for mitigation.[15]

Climate Change and Energy Scarcity as Fatalist Denial

It is easy to see that climate change and energy scarcity are escalating problems. Initially, a decline of energy supply by a few percent per year will simply trigger an economic crisis, but if the decline of energy supply continues for a decade or more then the ripple effects will be enormous. The same applies to runaway climate change. Initially, a few hot summers are not a huge problem. They may even be welcome to urban populations who prefer sunshine over rain. In the long run, however, a rise in global mean temperatures leads to mounting problems such as diminishing agricultural productivity, intermittent access to fresh water, and increased flood risks. This is further aggravated by the higher frequency of extreme weather events (see chapter 2).

It is less obvious that climate change and energy scarcity are also intractable, but many people operate under that assumption. Another way of saying that a problem is intractable is to call it a predicament. Unlike a tractable problem, a predicament per definition cannot be solved (Chisholm 1995). At best, it can be honestly acknowledged and somewhat mitigated. In principle, the same applies to the transitory nature of industrial society, or "human predicament." However, acknowledging this would require a moral rigor that is usually thwarted by fatalism and denial.[16] As we have seen, when intractable problems escalate, the best thing myopic individuals and groups feel they can do is to reduce their suffering at any given point in time. Both individually and as groups, people have a predisposition to minimize such pain by engaging in fatalist denial.

Due to their highly disquieting nature, planetary problems such as climate change and energy scarcity are the perfect breeding ground for fatalist denial. Because acknowledging them arouses high levels of emotional distress, denying their existence is an understandable and partly rational coping strategy to deal with the existential angst they are otherwise bound to engender. While it is extremely unpleasant to consider scenarios like those depicted in chapters 3 and 4, it is more congenial to ignore the problems and think of something more positive.

In a way, it is a déjà vu. The debate about limits to growth started in the early 1970s (Meadows et al. 1972) and culminated during the Carter administration, when there was a public inquiry into the issue (Council on Environmental Quality and US Department of State 1980). By the mid-1980s, when it had become clear that radically altering the industrial way of life was politically impossible, the debate ushered into denial with President Reagan (1985) solemnly declaring: "There are no limits to growth and human progress when men and women are free to follow their dreams."

Faithful to this somnambulistic doctrine, ever since the debate was silenced in the 1980s the limits to growth have remained unacknowledged.

Humankind is in denial of its existential predicament, and concerns about resource scarcity have largely led a fringe existence in public consciousness. There is an informal coalition between those who genuinely believe that there are no physical limits to growth and those who, at the bottom of their hearts, feel that all is not well but have opted for fatalist denial. In a way, those warning of peak oil have attempted a revival of the debate about limits to growth, but so far they have made only minor inroads into the dominant culture of denial.

As we have seen, fatalist denial not only reduces cost but also buys time. It can perhaps postpone the oath of disclosure of world industrial civilization and thus increase its life span by a few years or even decades. Business as usual can go on for a little longer, before climate change and energy scarcity come to a head. But the later the moment of reckoning occurs and the later the limits to growth are acknowledged, the worse climate change is going to be and the more drastically energy consumption will have to be curtailed. This is in line with the observation that, despite the twin benefits of reducing discomfort and buying time, fatalist denial leaves people with a greater mess.

Another downside of fatalist denial is that, while the fallout from intractable escalations can be mitigated, deniers conceal from themselves essential information. Having removed the problem from their consciousness, they are bound to miss any opportunities to cut their losses. Thus, in the case of climate change, industrial society foregoes the opportunity to preemptively adapt to some of the predictable effects of global warming. In the case of energy scarcity, it misses the opportunity to smoothen the descent.

Fateful Denial

The difference between fatalist and fateful denial is that, in the former, an escalating problem is intractable whereas in the latter it can, or could, be solved. It is a characteristic of escalating problems that, as the problem gets worse, possible solutions also become more difficult, and finally impossible. Therefore, the denial of escalating tractable problems can have pernicious consequences, since it obfuscates possible solutions until it is too late.

Let us imagine a man who is fearful of suffering from a serious progressing illness. He goes to the doctor and is diagnosed with lung cancer.

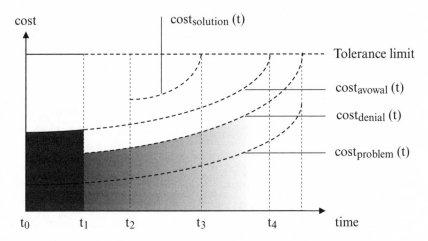

Figure 6.4
Fateful denial

At this point (t_1), he enters denial because he cannot cope with the anxiety of having a terminal illness. A few weeks later (t_2), the doctor calls him at home and points out that, in his case, the disease need not be fatal. There is a painful operation that is very likely to succeed, but only if he undertakes surgery in the next couple of months. After that (t_3), it will be too late for an intervention, and the disease will enter its terminal stage. Will our man stay in denial about the fact that he has lung cancer, or will he undertake surgery to save his life?

The situation is illustrated by figure 6.4. In the model, there is a window of opportunity between t_2 and t_3 to solve the problem, and thus avoid an impending disaster. Far-sighted people acknowledging the situation would certainly solve the problem during that time. After all, the cost of the solution is only momentary while the situation, if unchecked, endures and increases to catastrophic proportions. Only a myopic fool would procrastinate until the point when the cost of the solution exceeds the tolerance limit (t_3), and it is too late.

If there is denial, however, there is a real danger that the window of opportunity between t_2 and t_3 will be lost. This is for two reasons: the numbing effect of denial, and the greater likelihood of procrastination.

First: the numbing effect of denial. As we have already seen, regimes of denial are unconscious or half-conscious regimes of silence. Usually the deniers are only dimly aware of their problem. They hide the true cost of the situation from themselves. They may not even have accurate knowledge of their own utility functions. By implication, deniers limit their own access to information that might challenge the status quo and help them overcome their unacknowledged problem. This makes it impossible for them to access and freely exchange information about possible solutions. As a consequence, the availability of a solution will be obfuscated. Even an effectively resolvable situation is then unlikely to be solved.

Second: the greater likelihood of procrastination. Deniers are not stupid. Somewhere deep down, they know that there is a problem. But even when they become aware of a possible solution, they are more likely than others to miss the train because denial reduces cost, and thus the incentive to tackle the issue. To see this, let us consider procrastination (Akerlof 1991). From a long-term perspective, a problem should be tackled when the benefit of getting rid of the problem outweighs the cost of the solution. Usually this is the case, because the cost of the solution is a one-off while the problem, if not solved, will continue to cause pain. From a short-term viewpoint, however, it is often tempting to postpone a painful solution to tomorrow rather than tackle the issue today. As a consequence, even reasonable people have a myopic tendency to procrastinate.[17] Based on these considerations, it is easy to see why procrastination is more likely when there is denial. Problem solving becomes less urgent once a regime of denial has been established because, insofar as denial is less costly than avowal, the incentives for tackling the problem have been reduced.[18]

In sum, denial may take a possible solution to an escalating problem off the table. This may lead to fateful consequences that could have been averted in the absence of denial. For example, a cancer patient who could have been healed through a painful operation may die from his disease.

Climate Change and Energy Scarcity as Fateful Denial

While it seems obvious that climate change and energy scarcity are escalating problems, there is room for reasonable disagreement as to whether they are tractable or intractable. Personally I believe that they are largely intractable, and it seems to me that we have already passed the point of no return. Arguably, CO_2 emissions have already reached a level where runaway climate change cannot be averted any more (Hamilton 2010).

Moreover, in the current climate of financial instability it appears likely that openly admitting the looming risk of energy scarcity would be tantamount to committing suicide for fear of death. How would markets and investors react to public acknowledgment that the availability of energy is declining for good (Korowicz 2010)?

But many proponents of renewable energy and climate change mitigation operate under the assumption that we do not (yet) need to accept climate change and energy scarcity as humanity's inescapable fate. For example, the advocates of cap-and-trade see climate change as a market failure that can be addressed. Similarly, the IEA often calls for high levels of investment in renewable energy, oil exploration, and the extraction of non-conventional fuels to avert energy scarcity for the foreseeable future (see, for example, IEA 2012a, 2012b).

Ultimately, this is a judgment call. If climate change and energy scarcity are intractable, their disavowal falls under the rubric of fatalist denial. If however they are tractable, their denial must count as fateful. There is much to be said against fatalist denial, but the consequences of denial are even worse if we assume that climate change and energy scarcity are tractable problems. Indeed, if there is only a remote chance that climate change and energy scarcity are still tractable, failure to own up to these problems and act upon them would be a terrible sin of omission. In such a situation, fateful denial would contribute to making a resolvable situation entirely hopeless.

But if we have already passed the point of no return (t_2 in figure 6.3, t_3 or t_4 in figure 6.4), the question of whether we are in fatalist or in fateful denial does not matter. Once it is too late, there is no rational alternative to denial and self-deception. It then is a purely ethical question whether we prefer to live in the truth or in denial.

After Copenhagen

The distinction between fatalist and fateful denial can also help us understand the revival of climate skepticism and climate change denial after the failure of the 2009 Copenhagen summit. Apparently, the Copenhagen fiasco led to a shift from the perception of anthropogenic warming as a difficult but tractable problem to its perception as an existential predicament, followed by a tacit shift from attempts at climate change mitigation to climate change denial.

As long as there seemed to be a possibility, however remote, for an international deal to be reached, there was a critical mass of people pushing for vigorous action. In the run-up to the conference, stakeholders operated under the assumption that climate change is a tractable problem. For a few weeks and months, Copenhagen had become "Hopenhagen" (Death 2011).

After the fiasco, decision makers and the public lost confidence that climate change is tractable. Most people now feel that climate change is an existential predicament that cannot be solved. A further increase in the concentration of atmospheric carbon is seen as inevitable—and is happening. Since people tend to operate under the premise that lamenting intractable problems is pointless, climate skeptics have become emboldened (see chapter 5) and climate change denial is back with a vengeance (Hamilton 2010; Norgaard 2011).

The failure of the Copenhagen summit, then, has led to a radical mood swing, with the public now displaying a strong cognitive and emotional bias against "negative" worldviews. People have an understandable desire to focus their attention away from climate change, thus avoiding the considerable anxiety that is caused by acknowledging the problem. Insofar as climate change is subliminally seen as intractable, fatalist denial offers itself as a reasonably rational coping strategy ("If there is nothing we can do . . .").

In this situation, even the best efforts to galvanize awareness are bound to fail because the psychosocial mechanism of denial has kicked (back) in. From its inception, the IPCC has seen a heroic effort by scientists to engage all relevant stakeholders. But when it seems that there is nothing to be done, people have little patience with scientists and activists telling them to face the problem. Some will no doubt continue the fight, but decision makers are unlikely to follow because in a democracy they must worry about reelection.

In an atmosphere of fatalist denial, the incentives for decision makers are set in such a way that costly policies to address climate change appear impossible. For example, US president Barack Obama silently gave up his plan to introduce an unpopular cap-and-trade program for CO_2 emissions. Although Obama was personally committed to the plan, he must have felt that pursuing it any further would have been politically suicidal (remember Jimmy Carter?).[19] During the 2012 US presidential campaign, both Obama and his Republican contender Mitt Romney avoided the topic of climate change, and only the day after his reelection did Obama dare to mention climate change in his acceptance speech.

What Can Be Done?

Let us be optimists for a moment and assume that it is not too late. Let us imagine that we are climate scientists or activists, and that we desire to break the regime of denial in which people are caught. This raises the issue of social intervention: What can be done when denial does more harm than good?

Despite the harmful side effects, deniers themselves are mostly unable to end their denial because, ostensibly, they do not even know that there is a problem in the first place. Deep down, of course, they know that something is wrong. However, except in those instances where a situation breaks down under its own weight, it normally takes a social intervention to end denial.[20]

Let me propose a strategy for social intervention that logically flows from my models and then contrast it with other well-intended strategies that are less likely to succeed because they disregard the fact that denial is a quasi-rational strategy of pain avoidance.

My recommended strategy is rational persuasion: convincing deniers that it is in their best interest to admit their problems. Rational persuasion is premised on two insights. The first insight is that most deniers will acknowledge their problems only when the practice of denial is exposed as counterproductive. The second insight is that this will happen only if there is a sea change. Deniers need to be told plausible stories about the situation they are in and the counterproductive effects of their practice. In fact, it is sometimes possible to "renegotiate" the way deniers understand their situation. When deniers accept an alternative framing whereby denial is counterproductive, then an end to their denial is close at hand.[21]

If we apply these considerations to our models, then rational persuasion is most likely to succeed if the following prescriptions are followed:

1. Deniers should be persuaded that their orientation ought to be long term rather than short term. As we have seen, denial is tempting as a short-term strategy, but often has pitfalls in the longer term. It is therefore

necessary to turn the gaze of deniers to the long-term consequences of their failure to acknowledge the problem.

2. Deniers should be persuaded that their problem is getting worse. As we have seen, the denial of permanent problems can be rational, not only in the short term but also in the long run. Only escalating problems are scary enough to awaken deniers from their slumber.

3. Deniers should be persuaded that their problem, painful as it may be, can be solved. The best way to do this is to propose a specific solution. When a problem is intractable, deniers have a point; acknowledging the problem may not be worth the headache. But when a problem can be solved, they only damage themselves by denying it.

4. To counteract the risk of procrastination, deniers should be persuaded that, unless they act quickly, it will be too late because the solution will not be available any more.

In short, rational persuasion can be effective when deniers are persuaded that they should see their problem from a long-term perspective, that they are facing an escalating problem, and that an effective solution is possible—but only if they act quickly, because otherwise it will be too late. Thus, the best way to create the last-minute panic necessary to galvanize deniers into acknowledgment and action is to persuade them that their situation resembles the one described under the rubric of fateful denial (and that, in the process outlined in figure 6.4, they are somewhere between t_2 and t_3).

Funny enough, we have just reinvented the wheel of post-normal climate science (see chapter 5). First, climate scientists have developed scenarios projecting climate change far into the future, thus extending people's time horizons. Second, they have used dramatic visualizations of growing CO_2 emissions, such as the famous "hockey stick" diagram, to convince people that climate change is an escalating problem (Mann 2012). Third, they have contributed to specific policy prescriptions such as carbon reduction targets. Fourth, they have warned that in the absence of an adequate response there is a point of no return.

Climate scientists have done everything in their power to raise awareness of climate change as a vicious but tractable problem, and to promote vigorous action. They have worked for decades to convince decision makers and the public that the problem of climate change needs to be acknowledged and acted upon. And yet, there has been a fierce backlash. Precisely because denial is so tempting, even the most diligent effort to raise awareness and galvanize action is not immune to the risk that people will not listen.

Apart from rational persuasion, there are other methods of social intervention; but they all have serious downsides. For example, environmental activists have tried to create an atmosphere in which climate change deniers should feel ostracized, while those believing in climate change should feel in sync with the "do-gooder" mainstream. They have tried to persuade deniers that there will be psychological or social benefits from adopting an unvarnished view of the situation. However, this clashes with the laws of social gravity. When it comes to mobilizing social sentiment, the game is tilted in favor of those who tell people that all is well—and anyone who says otherwise is labeled a scaremonger.

An even worse strategy is to make the problem appear more menacing. Some people believe that deniers will immediately acknowledge their problem once they understand its gravity. Insofar as denial is in essence a strategy of pain avoidance, however, this cannot work. On the contrary, the incentives for denial are bound to increase as the situation feels more menacing. For example, many have been scared to death by the apocalyptic movie *The Day After Tomorrow*. After viewing the film, however, people were actually less likely to acknowledge climate change than before (Lowe et al. 2006, 443).

Yet another doubtful method is moral proselytism. Although proselytism has a bad press these days, it springs only from the best motives. From a Christian viewpoint, for example, it is wrong for people to deny their sinfulness. People should humble themselves and hope to obtain forgiveness by repenting of their sin. They should acknowledge that God only knows their wicked hearts, and only his grace can redeem them (Jeremiah 17:9–10; Dyke 1614). Despite the noble motives, most unconverted people see religious proselytizers as intrusive and prevent them from trying to save their souls.

A similar situation exists for non-religious proselytizers, or "truth apostles." Idealists of all stripes agree that only a life in truth is a life worth living (Havel 1989; Rowe 2010). Denial is dishonest and objectionable. Elephants in the room must be exposed (Zerubavel 2006). There is something respectable about this moral impetus to convert people to "inconvenient truths" (Gore 2006). Nevertheless, deniers are not amused. They sometimes even persecute truth apostles, for example by mobilizing opprobrium against "grumblers."

Once again, there is a simple utilitarian reason why moral proselytism is successful only in rare cases: denial is inherently about pain avoidance. Insofar as people have a tendency to minimize pain in the short term, they prefer comfortable lies to inconvenient truths. As many would-be prophets[22] and truth apostles have had to find out the hard way, deniers often lash back when their self-deceit is challenged by people "speaking the truth."

To be sure, it is sometimes possible to end denial by being disruptive. Deniers can be ambushed and cornered into admitting their problems. The downside of this crusading approach is that the cure may be worse than the disease. At least, this is what was famously discovered by the nineteenth-century Norwegian playwright Henrik Ibsen. In a period when Victorian morality was holding sway in Europe, small and intimate groups such as petty bourgeois families were the ideal breeding ground for denial. Ibsen exposed the realities behind the charades typical of such families. At the same time, he demonstrated that denial was integral to the very fabric of these families and the happiness of their members. In *The Wild Duck* (Ibsen 2006 [1884], 117–118), a worldly and wise doctor states: "You take away the life-lie from the average man, and you take away his happiness with it."[23]

As an afterthought, and to state the obvious: when people are in denial, it cuts no ice to offer them "objective" information in a dispassionate way. It is simply not true that deniers would acknowledge their problem and act upon it "if only they knew." While this may happen in situations where there is genuine ignorance, information campaigns are doomed in situations where people have clandestine reasons for not wanting to know.

Bad Things Go Together

The moral economy of inaction has four elements: temporal discounting, spatial discounting, collective action problems, and denial. The combined effect of these elements is that our collective failure to confront the limits to the current growth model is causally overdetermined. This becomes clear if we apply counterfactual thinking: There is no single causal factor in the absence of which people would suddenly confront the transitory nature of industrial society. For example, take denial away, and discounting as well as collective action problems still remain.

As if this were not bad enough, the elements are also interconnected in vicious ways. For example, the likelihood of denial is increased by the presence of ethical discounting. People have an incentive to deny problems that happen far away not only because the consequences are suffered by distant strangers but also because such problems are less likely to appear tractable to them. People are also more likely to deny problems that appear to be distant in time, as temporal discounting suggests that such problems can be ignored with relative impunity.

Collective action problems also make denial more likely. For example, it is rational for me as a private citizen to deny climate change while tacitly hoping that others will deal with the issue (free riding). By the same token, it may not be smart for you and me as self-interested individuals to acknowledge collective problems like climate change and the risk of energy scarcity unless enough others do the same (prisoner's dilemma).

Bottom Line

All of this confirms insights from evolutionary psychology. Arguably, "mind your own business" was an adaptive strategy prior to the era of planetary risks. For most of human history, it made a lot of sense to discount the future and distant strangers. Collective action problems were effectively solved at the communal level. Anything beyond that level did not count for much—except for a thin stratum of aristocrats, merchants, and priests. For most commoners, problems beyond the immediate challenges of everyday life were largely intractable, so ignoring or denying them was perfectly reasonable. The last few centuries if not millennia have significantly reshaped human culture. Even the least developed parts of the world are now highly interconnected with what many see as an emerging world society. Nevertheless, given our common evolutionary heritage, it is hardly surprising that people have no time for planetary risks such as climate change and energy scarcity.

7

Where to Go from Here

Most people act like the inductivist turkey in chapter 1. They trust that what has sustained our prosperity and growth in the past will continue to do so. Even ecologically sensitive individuals and groups focus on mitigating damage caused by industrial society, rather than confronting the disconcerting fact that industrial society as such is the least sustainable form of civilization in history.

Current policies and political discourse are thus paradoxically geared toward "sustaining the unsustainable" (Blühdorn 2010, 2011). There is no appropriate strategic governance of long-term risks to provide, in so far as possible, a softer landing when industrialism enters its terminal decline. This is deeply regrettable because, if we fail to confront the human predicament, we are likely to have an enormously hard landing ahead. And yet, few of our fellow turkeys are ready to give up the comfortable habit of denial. As we have seen in chapter 6, there are numerous reasons why even well-informed people tend to be neither able nor willing to seriously confront the transitory nature of industrial society.

If this is so, then why revolt against the human predicament? Why bother, if persuading people of the transitory nature of industrial society is as impossible as persuading turkeys to vote for Christmas—or Thanksgiving, for that matter?

Here is my personal response: The best thing a moral individual can do is to try to live "in the truth." Life is tragic and sometimes there are no solutions. Not every disease can be cured. Insofar as climate change and energy scarcity are part of the human predicament, even the most accurate diagnosis is unlikely to suggest an easy cure. And yet, my mission as a scholar is to get to the bottom of things regardless of whether or not there is a solution. This does not mean that, as a citizen

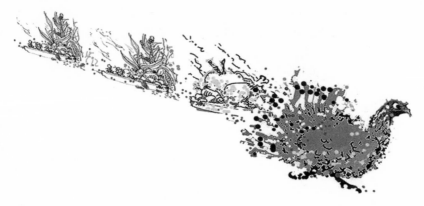

Figure 7.1
The denialist turkey. *Source:* Courtesy of Horst Friedrichs.

and consumer, I am better than anyone else. My task as a scholar is not to save the planet or pose as an ecological do-gooder. It is plain old-fashioned intellectual honesty.

Some readers will find this depressing. My attitude will sound weary to those believing that problems like climate change and energy scarcity can and must be dealt with either through politics (Heinberg 2006; Giddens 2009) or local activism (Hopkins 2008; Murphy 2008; De Young and Princen 2012). It will sound entirely outrageous to those setting their hope in a cornucopian can-do attitude and believing that aspirational statements and positive thinking can revolutionize what is politically feasible (Nordhaus and Shellenberger 2009).

Of course, I would be delighted to be proven wrong. In the next few pages I will make an effort to be more constructive and ponder what it would take to confront the current sustainability crisis. My suggestion is that, at least in theory and in an ideal world, the best chances for me to be proven wrong would stem from a combination of *resilience thinking* and *ontological securitization*.

Resilience Thinking

Resilience is the ability to bounce back from a personal or systemic crisis, altered perhaps but essentially unscathed. In general terms, it is the capacity of a system to "continually change and adapt yet remain within critical thresholds" (Folke et al. 2010, 1). Or, more specifically, it is the ability of a system to "absorb a spectrum of shocks or perturbations and to sustain and develop its fundamental function, structure, identity and feedbacks as a result of recovery or reorganization in a new context" (Chapin III et al. 2009, 241).

Under normal circumstances, a system is a conglomerate of largely autonomous subsidiary systems, or subsystems. Every subsystem follows its own logic and continuously adapts to a myriad of perturbations and challenges. But sometimes this is not enough. In a general system crisis, the forgotten hierarchy of functions and values comes to the fore. It turns out that the subsystems are not an end in themselves but serve essential top-level system functions and values. These top-level system functions and values provide purpose to the subsystems and constitute the framework under which they operate. They must be defended even at the price of shaking up the subsystems and their secondary values. In a general system crisis, all subsidiary systems must be up for grabs because there is an ultimate end: to secure the resilience of the system as a whole and the core functions and values enshrined in it.

This is called *transformability* and can be defined as "the capacity to create untried beginnings from which to evolve a new way of living" (Folke et al. 2011, 724).[1] Something like this occurs, for example, in the metamorphosis of an insect. It is what the caterpillar experiences when turning into a butterfly. From a social-ecological, socio-metabolic, or earth system perspective it must also occur in the next transition of our civilization (Berkes et al. 2003; Walker et al. 2004; Haberl et al. 2011; Bierman et al. 2012).

In that case, the top-level system function is the maintenance of the relatively stable climate within which human civilization has evolved over the past 10,000 years. The so-called Holocene climate is "the only state of the Earth System that we know for sure can support contemporary society" (Steffen et al. 2011, 739). It is the linchpin of humanity's life-support system and stands at the core of its safe operating space (Rockström et al. 2009). It can even be argued that keeping the Holocene climate resilient is an essential system function not just for human society but for the planetary system as a whole (Lovelock 2000).

Regardless of its vital importance, climate stewardship must not come at the cost of the core values that make our existence worth living. Apart from the Holocene climate, we must also preserve our core civilizational values. But what are these values? In my view (not derived from any firm literature), there are two of them: our ability to shape the environment, and our humanity in the most general ethical sense of human goodness. It seems to me that there is little intrinsic value to human civilization without these two.

Take away our ability to shape the environment, and human civilization becomes meaningless. A degree of control over the environment is what distinguishes civilizations from hunter-gatherer societies, and we clearly cannot afford to let it go. Imagine what would happen to the world if billions of people were forced to revert to a foraging lifestyle! This is of course not to suggest in any way that civilizations are good environmental stewards. They have mostly been very bad stewards, and yet stewards they have been. Even bad stewards have a certain degree of control over their environment, and if anything this degree of control has steadily increased since our departure from Eden. It is not by accident that Nobel Prize–winner Paul Crutzen has popularized the term *Anthropocene* for the current human-controlled geological era following the Holocene.

For all practical purposes, our ability to shape the environment depends on our ability to harness energy. Energy inputs are not the only thing needed to shape our environment (we also need material resources and knowledge), but without them it would be impossible to maintain the considerable control of nature that we have achieved. Because this is so, a viable energy regime is as fundamental for the maintenance of civilization as the preservation of the Holocene climate.

The other core civilizational value is our humanity, or "goodness." Once again, this is not to suggest that human civilizations have ever been particularly "humane." Civilization has brought the worst and the best out of humans, and our humanity has often been eclipsed by terrible atrocities. And yet, after every civilizational crisis, human goodness in its manifold forms has reemerged and in that sense proven resilient. We need to make sure that it remains that way.

That is not a trivial undertaking, as can be seen from the case studies discussed in chapters 3 and 4. The imperial Japanese and North Korean reactions to the threat of energy scarcity suggest that fundamental human values are easily jettisoned when civilization's ability to harness energy comes under jeopardy. By the same token, the history of the ancient Near

East and the medieval Far North suggests that terrible atrocities must have occurred in situations of severe eco-scarcity resulting from climatic stresses (in some cases including "dark ages"). Our humanity remains a fragile and tenuous accomplishment that must be made resilient against the looming climate and energy crisis.

Given the magnitude of the challenge, climate stability and the core civilizational values of environmental control and human goodness cannot be secured without the transformability of subsidiary systems. Radical transformations of our political, legal, and economic subsystems are necessary to prevent the degradation of our climatic life-support system and core civilizational values. Transformability is needed at multiple levels, from individual citizens through local communities and nation-states up to the global level, including industrial civilization as a social-ecological system (Folke et al. 2011). In the parlance of resilience thinking, "deliberate transformation involves breaking down the resilience of the old and building the resilience of the new" (Folke et al. 2010, 7).

The problem is that resilience is not always and everywhere a good thing. The resilience of the subsystems can be a serious obstacle to the transformability and thus to the resilience of the system as a whole. In socio-ecological transformations, the "resilience of behavioral patterns in society is notoriously large" (Folke et al. 2010, 2). Reforming political, legal, and economic subsystems is difficult because of various forms of lock-in. We all know how entrenched these subsystems are and how hard it is to reform even minute aspects of them. The resilience or recalcitrance of the subsystems can become a serious obstacle when the system as a whole is at stake, which may lead to tragic outcomes.

In sum, there are three top-level system functions that need to be, or be made, resilient in the face of planetary challenges: the Holocene climate, our ability to shape the environment, and our humanity. Everything else must be subordinated to these. If we do not safeguard them, instead of a voluntary transformation we will face an involuntary one that will put an end to civilization as we know it.

Ontological Securitization

The stabilization of the Holocene climate and the preservation of our humanity and ability to shape the environment are literally questions of "to be or not to be." Insofar as climate change and energy scarcity constitute a general system crisis, we must take radical measures because in the absence of a system transformation everything else will be lost. The only way the recalcitrance of the subsystems can be broken is to define the preservation of top-level system functions and values as a security issue in the most existential, or *ontological*, sense.

Therefore, what is required is an ontological form of securitization. *Ontological security* is "the confidence that most human beings have in the continuity of their self-identity and in the constancy of the surrounding social and material environments of action" (Giddens 1990, 92). *Securitization* is the act of declaring that something is a security issue because essential values are at stake, and that emergency measures must be taken lest these values be lost (Buzan, Wæver, and Wilde 1998). Hence, *ontological securitization* is the act of declaring that emergency measures are needed to prevent a loss of our self-identity and to preserve the constancy of the environments in which we are able to act.[2]

When ontological security and basic trust are shaken, existential anxiety kicks in: "On the other side of what might appear to be quite trivial aspects of day-to-day action and discourse, chaos lurks. And this chaos is not just disorganisation, but the loss of a sense of the very reality of things and other persons" (Giddens 1991, 36). "Why is everyone not always in a state of high ontological insecurity, given the enormity of . . . potential existential troubles?" (Giddens 1990, 94).

The answer is, of course, denial. When acknowledged, predicaments such as climate change and fuel depletion arouse tremendous anxiety. As we have seen in chapter 6, denial is a typical reaction to existential trouble. People evade ontological insecurity by ignoring their problems. But, as we have also seen, the consequences can be tragic because denial often makes a bad situation worse and prevents possible solutions from being adopted while there is time.

Ontological securitization is, or would be, a more proactive approach to confront the current sustainability crisis. It represents the attempt to overcome the dread connected with ontological insecurity by gaining a heightened sense of agency. For this to happen, climate change and fuel depletion must be correctly identified as the actual source of our ontological insecurity.

As was argued in the last section, the values at stake are climatic stability and our core civilizational values. Tragically, however, securitization moves usually either do not occur due to denial or they are diversionary rather than targeted to the real problems. Precisely because conscious awareness of ontological insecurity is so dreadful, securitization typically does not address the deeper existential issues but rather diverts attention to some sideshow where it appears possible to "do something" without confronting the actual source of ontological insecurity.

Examples abound. During the 2008 US presidential campaign, conservative Republicans reduced energy insecurity to dependency on foreign oil imports and promoted environmentally harmful extraction ("Drill, baby, drill") as a solution.[3] Another familiar strategy is to depict energy insecurity as the consequence of resource nationalism in developing countries, combined with a call for forcible regime change. Yet another diversionary move is to call for carbon capture and storage (CCS) or geoengineering to secure the continued reliance on fossil fuels.

The problem with diversionary securitization is that, given the real problems facing humankind, it is delusionary. It must be understood as a strategy of self-deception to flank denial about the real underlying problems.

Insofar as our climatic life-support system and our core values are at stake, it is counterproductive and irresponsible to securitize secondary aspects of the impasse such as this or that nation's access to cheap energy. The only adequate way to deal with planetary challenges such as climate change and energy scarcity would be ontological securitization at the level of earth system governance and core civilizational values. What is required, then, are securitization moves that correctly diagnose the sources of our ontological insecurity and suggest viable steps toward a sustainability transformation, hopefully including a new social contract to democratically shape the transition (WBGU 2011).

In this context, it is necessary to question an assumption of conventional securitization theory: namely that the emergency measures resulting from securitization automatically imply a suspension of the democratic process (Buzan, Wæver, and Wilde 1998). This is certainly what happens in many cases, and not by accident (Schmitt 1932; Agamben 2005). In principle, however, there is also a civic republican way of dealing with an emergency by a combination of enlightened leadership and intensified public debate and participation (Pettit 1997). From such a civic republican viewpoint, ontological insecurity in the face of climate change and energy scarcity requires not only an intense awareness of the core values at stake but also a vibrant democratic debate about the nature of the challenge and the measures to be taken (Orr 2009; Barry 2012).

This debate must not only raise but also settle a number of fundamental questions with a view to preserving our climatic life-support system and core civilizational values. Without prejudicing the answers, here are some examples.

To keep the global climate within Holocene conditions, shall we try to voluntarily reduce the human footprint or shall we strive for active planetary stewardship (Rockström et al. 2009; Steffen et al. 2011)? How can we make sure that the multiple linkages between social and ecological systems are adequately matched by governance arrangements at the local, national, regional, and global levels (Folke et al. 2011)? Under which conditions is social and technological innovation likely to promote the sustainability transition, and when is it more likely to reinforce unsustainable patterns of lock-in (Westley et al. 2011)?

Shall we rely on large-scale technological solutions, maybe including aggressive nuclear programs, to de-carbonize industrial society? Shall we undertake the equivalent of a war-time mobilization to promote renewable energy and environmentally friendly technologies (Spratt and Sutton 2008; Dyer 2010)? Shall we actively pursue a strategy of frugality and deliberately forego economic growth (Victor 2008; Jackson 2009)? Or, even more radically: Shall we pursue low-tech solutions at the local level and seek our refuge in communal "lifeboats" (Hopkins 2008; Greer 2008; De Young and Princen 2012)? And how can we safeguard basic values under conditions of scarcity (Ophuls 1977, 2011)?

Such vital questions must be openly and intrepidly discussed in the public domain. The debate must happen at all levels of governance, from local to national and from regional to global. From a pragmatic viewpoint, "adaptation decisions should . . . be taken such that they are robust to present circumstances and value systems and not ruled out by potential future states" (Adger, Brown, and Waters 2011, 699). The prize is not civic debate for its own sake but the formulation of policies to confront ontological insecurity by introducing democratically acceptable measures to mitigate the sustainability crisis.

Bottom Line

What a dream! Writing the last few pages has been a veritable feast, far more enjoyable and also somewhat easier than writing the rest of the book. My only problem is that, as a pessimist—or better, as a realist— I doubt whether the dream can come true.

As we have seen in chapter 1, industrial society is hopelessly in overshoot. As we have seen in chapter 2, climate change and energy scarcity are daunting problems that can perhaps be mitigated but hardly be averted. As we have seen in chapters 3 and 4, the social and political consequences will vary but must be expected to be very serious. As we have seen in chapter 5, existing knowledge regimes are inadequate. And as we have seen in chapter 6, there is an entire moral economy of inaction to prevent us from tackling the issues.

Is there any indication in the history of civilization that we are able to rule ourselves or each other to our own and/or the planet's benefit? Despite some encouraging examples at the local and communal level (Ostrom 1990), civilization's overall record in managing the environment is dismal (Wright 2004; Diamond 2005). History is littered with failed civilizations, sometimes followed by dark ages (Tainter 1988; Chew 2007, 2008). This is in line with broader speculation about a cosmic propensity toward growth in complexity, which however is ultimately constrained by the second law of thermodynamics (Chaisson 2001; Christian 2004; Kauffman 1995).

Many will find this depressing, and to cheer them up there is a variety of uplifting accounts of the situation we are in. For example, some follow the principle of hope and indulge in the fantasy that this time around we will get it right (Rifkin 2011; Kamal 2011; Brown 2011; Bryner 2012; Hoffmann 2012). Others call for positive thinking and declare that pessimism does not flow from the situation but actually contributes to the impasse (Nordhaus and Shellenberger 2009). Yet others offer a carefully calibrated balance between hope and despair to hedge their bets (Brown and Sovacool 2011). While there is time to talk, the debate continues.

But is it reasonable to assume that humankind will be able to redeem itself, after all that we know about human nature and the history of civilizations? To be sure, there are cases of civilizations that have moved to a higher level of complexity when an upgrade was available. But where are the examples of advanced civilizations that have been able to pull themselves back from the brink and voluntarily downgrade their complexity when an upgrade was not available?

Given the manifest inability of civilizations to voluntarily reduce their complexity, it seems equally if not more plausible to expect a long emergency preceding a distant eco-technic future (Kunstler 2005; Greer 2009), or that one fine day God will "destroy those who destroy the earth" and "make all things new," providing "a new heaven and a new earth" (Revelation 11:18; 21:1; 21:5).

Notes

Acknowledgments

1. Chapter 4 is an updated and extended version of my article in *Energy Policy* (2010). Portions of chapters 5 and 6 go back to my article in *Futures* (2011). Part of chapter 6 is drawn from my article in *Philosophical Psychology* (forthcoming).

Chapter 1

1. The story is common lore, but the source seems to be Russell (1912, 97–98).

2. For an introduction to the concept of carrying capacity, see Cohen (1995).

3. This results from an application of the so-called rule of 70: to estimate how many years it takes for something to double, take the number 70 and divide it by the growth rate.

4. This is because a century amounts to more than four doubling times of twenty-three years each.

5. Mathematical proof: $(1.03)^{100} \times 5.2\% = 1$.

6. For another warning, see Ehrlich and Ehrlich (2004). See also Bardi (2011).

7. So far, the original standard run scenario of 1972 is largely on track with historical data (Turner 2008; Hall and Day 2009).

8. This refers to the 2004 version. In the original version (1972, 124), the projected contraction of world population by 2010 was "only" to the level of about 1980.

9. Personal communication from the Project Manager of the Global Footprint Network, March 15, 2010.

10. In theory, political regulation to increase efficiency addresses the problems of resource depletion and environmental pollution at the same time. In practice, however, the benefits from efficiency gains are often thwarted by the fact that improved efficiency lowers cost and thus encourages increased consumption (the so-called rebound effect, or Jevons paradox; see Sorrell 2007).

11. Note that key sectors of the service economy, such as healthcare and information technology, heavily rely on an energy-intensive industrial base: pharmaceutical industry, computer industry, etc.

12. For a selection of related books, see Randers (2012); Klare (2012); Brown and Sovacool (2011); Moriarty and Honnery (2011); Heinberg and Lerch (2010); Homer-Dixon (2009); Smil (2008b); Taylor (2008); Gautier (2008); Homer-Dixon (2006).

Chapter 2

1. The wide range of the projections is mainly due to the divergence of climate models and emission scenarios: while the former are based on different assumptions about the climate system, the latter make different assumptions about future economic and political trajectories.

2. For an imaginative though sensationalist account, see Lynas (2007).

3. The uncertainties are high and the geographical resolution of the models is low. This is unfortunate because precipitation regimes can vary greatly over relatively small geographic distances, for example from one valley to the next.

4. Similarly, some of the IPCC's emissions scenarios consider the possibility of reduced international trade and cooperation, but none of them considers the eventuality of a terminal decline of industrial civilization (nor do they consider the adoption of any deliberate policies to mitigate climate change).

5. *Source:* http://www.esrl.noaa.gov/gmd/ccgg/trends.

6. 1980s: 2.0 percent per year; 1990s: 1.0 percent (*source:* Peters et al. 2012, 3).

7. Assumptions about energy efficiency are equally bold, with improvements in energy efficiency assumed to contribute more than half of all carbon abatement (IEA 2012a, 241). The following statement is worth quoting in full: "Over the Outlook period, energy intensity—energy demand per unit of GDP—declines by 2.4% per year and CO_2 intensity—CO_2 emissions per unit of energy use—falls by 1.8%. *When compared to recent trends, the challenge of the 450 Scenario becomes clear. Over the last ten years, energy intensity declined by only 0.5% per year, while CO_2 intensity grew by 0.1% per year*" (IEA 2012a, 252; my emphasis). For a particularly upbeat assessment of energy intensity, see Rühl et al. (2012).

8. Alarmingly, the downward trend was temporarily reversed in 2010.

9. Most of this investment would have to be mobilized after 2020, as until then the international community is for all practical purposes very unlikely to be more ambitious than assumed under the IEA's New Policy Scenario.

10. An exception is artificial cooling by planting brighter crops and whitening roofs and roads, because the measure (1) is reversible and therefore can be adopted by trial and error; and (2) has primarily local effects; but its impact is equally low.

11. The concept of peak oil was introduced by Colin Campbell and Jean Laherrère (1998), based on pioneering work by Marion King Hubbert (1956, 1962, 1981, 1993). More recent contributions to the debate are legion and cannot all be listed here (for select readings, see Heinberg 2003, 2004; Leggett 2005; Simmons 2005; Hirsch, Bezdek, and Wendling 2005; Hirsch 2008; UKERC 2009; Korowicz 2010; Owen, Inderwildi, and King 2010; Aleklett et al. 2010; Sorrell et al. 2010b; Fantazzini, Höök, and Angelantoni 2011; de Almeida and Silva 2011; Murphy and Hall 2011; Sorrell et al. 2012; Murray and King 2012).

12. Remember that the New Policy Scenario is consistent with a 3.6°C rise in global temperatures.

13. There is an optimistic estimate by Leonardo Maugeri (2012) of 2–3 percent, but most others including the IEA are less sanguine (see Kerr 2012).

14. For a selection of recent further readings on EROI, see Gagnon, Hall, and Brinker (2009); Gupta and Hall (2011); Guilford et al. (2011); as well as the speculative piece by Dale, Krumdieck, and Bodger (2011).

15. It is true that coal production has peaked in some countries and reserves data have been downgraded in recent decades. But this has been related to commercial rather than geological factors. While in Europe the "easy" coal may have been depleted, coal reserves in other places are still easy to exploit. It is reasonable to expect that, after a peak in oil production, demand for coal, and thereby recoverable coal reserves, will rise again. To the extent that coal becomes commercially more attractive, reserves data will then be readjusted upward.

16. In theory nuclear fusion would solve the resource problem once and for all, but in practice the technology is not available within the necessary timeframe.

17. The cut to zero is mainly premised on the following assumption: power production will rely on biofuels, and the carbon will be sent underground via CCS. Since plants sequester carbon during their growth and since that carbon is not re-emitted, the procedure takes carbon out of the atmosphere. This will then offset unavoidable CO_2 emissions from other economic activities. It sounds almost too good to be true, and in fact it may not be feasible on a large scale. What would happen to soil fertility if biomass were constantly withdrawn on a massive scale?

18. Capitalization analysis shows that, at least for the time being, financial markets are not betting on such an energy revolution (DiMuzio 2012).

19. For an overview of this and similar "escape routes," see van den Bergh (2012).

20. Especially the decline of petrol-based transportation after peak oil may pose a serious challenge to the implementation of ambitious modernization programs.

Chapter 3

1. The mandate of Working Group I covers the physical science of climate change. Working Group III deals with technological fixes to mitigate the effects.

2. Working Group II terms people as "human systems," denoting the human organism as a biological system. This must be distinguished from social and political systems.

3. For an overview, see Mearns and Norton (2010); Scheffran et al. (2012).

4. For recent surveys, see Bernauer, Böhmelt, and Koubi (2012); see also Mildner, Lauster, and Wodni (2011).

5. For a slightly modified version of the model, see Homer-Dixon (1999, 134); see also Kahl (2006, 59).

6. For examples of this quantitative literature, see Urdal (2005); Binningsbø, de Soysa, and Gleditsch (2007); see also Theisen (2008), undermining earlier findings by Hauge and Ellingsen (1998) in support of eco-scarcity theory.

7. There are indeed some lurid examples of environmental determinism combined with the politics of fear, such as the 2003 report to the Pentagon on the national security implications of climate change (Schwartz and Randall 2003). However, even environmental alarmists have mostly followed a more differentiated and multifaceted approach (Woodbridge 2004; CNA Corporation 2007; Campbell 2008; Herman Jr. and Treverton 2009; Dyer 2010; Welzer 2011).

8. The negative effects of climate change have weakened during the industrial era, but more so for the rich industrial countries of the North than for poor developing countries of the South. While the relationship between climate and conflict has largely disappeared in the North (Tol and Wagner 2010), in the South it remains strong and significant (Hsiang, Meng, and Cane 2011).

9. As the goal is "process tracing" to better understand the social and political effects of climate change, rather than a logical or statistical demonstration of causality, I am free to purposefully select my cases on both the dependent and independent variables (Collier, Mahoney, and Seawright 2004).

10. See the accounts by Kunstler (2005); Homer-Dixon (2006); Greer (2009).

11. This refers to the Levant and eastern Mediterranean, not to the global level.

12. The presence of suitable species of plants and animals was a necessary condition for this transition to happen. The condition was met in southwest Asia, as well as some other places where agriculture evolved independently.

13. For an earlier related theory, see Wilkinson (1973).

14. Confusingly, in most of the literature the northern outpost is called "Western Settlement" and the southern outpost "Eastern Settlement."

15. Unlike Greenlanders, Icelanders never had a taboo against eating fish.

16. This is not to deny the frequent starvation of individual Inuit hordes and the collapse of their settlements due to shifting migration patterns of wild animals.

17. See also the theoretical piece by Gort and Wall (1986) and the case study by Lang (2009) on diminishing returns on R&D in German manufacturing.

18. Even in information technology, gone are the days when a genius like Steve Jobs could generate major breakthroughs in Californian garages.

19. This is not to deny that industrial civilization may either be scavenged or reassembled into lower-complexity production systems (Greer 2009).

20. Most studies examine weather rather than climate (Gleditsch 2012, 7). They analyze the impact of short-term variations in rainfall and temperature on the incidence of violent conflict, and make the somewhat problematic assumption that a long-term shift in climatic conditions will have the same effect. For a typical example, see Theisen, Holtermann, and Buhaug (2012).

21. Modern industrial civilization has replaced territorial conquest by intensified resource extraction (first domestically, then internationally).

Chapter 4

1. The gruesome exception was the Pacific island of Henderson where people ended up cannibalizing each other to the last man after vital tools were no longer delivered from a neighboring archipelago (Diamond 2005, 120–135).

2. Minor cases not considered include isolated outposts such as West Berlin in the 1948 crisis and seriously embargoed populations such as the Palestinians in Gaza.

3. In 1993 China refused to step in for Russia, demanding hard currency for any further exports and radically cutting deliveries of "friendship grain."

4. Coal was used in the production of fertilizers both as an energy source and as a chemical feedstock (Williams, Hippel, and Nautilus Team 2002, 117–119). Fertilizer use fell by more than 80 percent from 1989 to 1998 (FAO/WFP 2008, 14).

5. See the Special Reports of the Crop and Food Security Assessment Mission to the DPRK (especially FAO/WFP 1999, 2008; WFP/FAO/UNICEF 2011).

6. Others have tried to explain the Great Famine by natural calamities (Woo-Cumings 2002). This was also the preferred explanation of the North Korean regime. But while Pyongyang has often tried to excuse the fiasco by citing floods and droughts, these were only the trigger.

7. For North Korea's chronicle of a death foretold see the books by Nicholas Eberstadt (1999; 2009, 275–312).

8. Official Cuban figures for the decline of imported raw materials and other vital inputs to industrial production and electricity generation were on a similar level (Wright 2009, 68). Even according to the most conservative estimate of the US Energy Information Administration, between 1989 and 1992 oil consumption in Cuba fell by 20 percent and the net consumption of electricity by 22 percent (http://www.eia.gov/countries).

9. In fact, Cuba is sometimes cited in the popular literature as a favorable contrast to North Korea (e.g., Pfeiffer 2006; Wen 2006).

10. To some extent, Cubans were helped in their efforts to cope with the crisis by a benign climate, remittances, foreign investment, and international aid.

11. See for example Rosset and Benjamin (1994); Altieri et al. (1999); Funes et al. (2002); Cruz and Sánchez Medina (2003).

12. Also, it is important to note that the Special Period had mixed effects on the environment; see Díaz Briquets and Pérez-López (2000).

13. See Moran and Russell (2009) on the "militarization of resource management."

14. For a long-term perspective see Greer (2008, 2009).

15. For a previous attempt, see Elhefnawy (2008).

16. For evidence on similar practices during the Cold War, see Lipschutz (1989).

17. Britain might try to evade the quandary by stressing its special relationship with the United States, but it is debatable whether the United Kingdom could offer the United States enough benefits to justify the burden of provisioning sixty million people with affordable fuel.

18. Despite an increasing internal market, China remains highly dependent on the exportation of manufactured products. It may not yet have accumulated enough economic wealth to insulate itself against the demise of international free trade.

Chapter 5

1. Full documentation is preserved in the online archive at issuecrawler.net.

2. Since the 1980s, the requirement of physically transferring oil stockpiles to undersupplied member states has been replaced by an obligation to release the same stockpiles on the open market to keep prices from skyrocketing.

3. The largest disruption was the 1979 Iranian revolution with a 5.6 percent shortfall, followed by the oil crisis of 1973–1974 and the 1990–1991 Gulf War with a 4.3 percent shortfall each (IEA 2012d).

4. In the 1991 Gulf War, the shortfall amounted to 4.3 percent. In the case of Hurricane Katrina, it was 1.5 percent. In the Libyan crisis, it was 1.6 percent (IEA 2012d).

5. In 1973, most net importers of oil, including the United States, were industrialized countries. Therefore, the OECD was appealing as an institutional hub to accommodate the IEA.

6. The links are with YouTube and CNN videos; and a report posted on Bloomberg.com about an oil price hike after an IEA demand forecast.

7. The appearance of normal science is due to the lack of open politicization.

8. This section benefits from a commissioned report by Chris Vernon who in March and April 2012 gathered information from the following insiders: Roger Bentley, Julian Darley, Aaron Dunlap, Adam Grubb, Nate Hagens, John Hemming, Rembrandt Koppelaar, Kyle Saunders, and David Strahan.

9. For an earth scientist opposing peak oil theory, see Gorelick (2010). Recently, however, even energy economists have come to acknowledge that world oil production has peaked and is now on a plateau (Yergin 2011).

10. Adam Grubb, email of March 22, 2012.

11. Julian Darley, interview on April 16, 2012.

12. Adam Grubb, email of March 22, 2012.

13. See http://www.theoildrum.com/node/5186. Data are no longer online, but some idea can be gleaned from the comments ("as one of the 7.2 % minority, my problem is getting my husband to see reality").

14. Rembrandt Koppelaar, email of April 16, 2012.

15. Chris Vernon, observation from attending several conferences.

16. Nate Hagens, email of March 26, 2012.

17. There are two links to commercial entities: BP, and the investment bank Simmons and Company. It is noteworthy that the founder of the latter, Matt Simmons, was a prominent peak oiler (Simmons 2005).

18. The 1987 *Montreal Protocol on Substances that Deplete the Ozone Layer* is often seen as a success story that should be, and indeed has been, emulated in the case of climate change (Litfin 1994; Canan and Reichman 2002).

19. For prominent examples of climate scientists sympathetic with climate alarmism, see Hansen (2009); Schneider (2009).

20. This point was anonymously made by a scientist peer reviewer of this book.

21. For concerned climate scientists, the precautionary principle would mandate action even if the probability of abrupt climate change were estimated as low. For climate skeptics, the burden of proof is reversed: if the scientific evidence is not 100 percent incontrovertible and agreed upon, no costly action is required.

Chapter 6

1. In line with common parlance, I reserve the term "denial" for the half-conscious or unconscious disavowal of a problem. Other terms, such as "climate skepticism," are used for deliberate disavowal (see chapter 5).

2. See Hume (1739, Vol. II, Part III, Section VII).

3. $9524 \times 1.05 = 10000$.

4. $1420 \times 1.05^{40} = 10000$.

5. Most economists assume that the growth-based time preference is equal to the real anticipated growth rate ($\eta = 1$), but it may carry a different weight. Essentially, η is a measure of how rapidly marginal utility is declining as wealth or the standard of living are increasing.

6. But see Weitzman (2012) for an economist coming close.

7. Classical assumptions of exponential discounting and temporal consistency are frequently too optimistic, and it often makes more sense to assume time inconsistency and hyperbolic discounting (Strotz 1955; Ainslie 2001, 24–27).

8. For a primer on collective action, see Olson (1965) or Hardin (1982).

9. Another fundamental problem is that, unlike communal grazing land, a sustainable climate is not easily amenable to privatization (i.e., "enclosed").

10. As the history of the Kyoto Protocol has shown, due to its limited signifi-
cance the European Union is not a credible candidate for climate leadership.

11. Portions of this section are taken from Friedrichs (2012).

12. In mathematical terms: $cost_{denial}(t) > cost_{avowal}(t)$.

13. This happens when: $\int_{t_1}^{t_3} cost_{denial}(t) > \int_{t_1}^{t_2} cost_{avowal}(t)$.

14. In mathematical terms: $cost_{problem}(t_3) > cost_{problem}(t_2)$.

15. Virtuosos in doublethink will awake from denial for a short time around t_2,
try some mitigation strategies, and then relapse into denial until the day of final
reckoning. This is difficult however, due to the numbing effect of denial.

16. Human mortality is the archetypical existential predicament leading to fatalism, sometimes of the defiant but mostly of the denialist variety.

17. The easiest way to see this is by contrasting entirely myopic and entirely
hyperopic behavior, as I have done here. It would be more realistic to consider
discount functions. However, this would not alter the basic finding.

18. There is an additional, closely related reason why procrastination is most
likely when people are in denial: procrastination is about myopic pain avoidance,
and deniers are the myopic pain avoiders par excellence.

19. The active efforts of climate skeptics and other interested parties to discredit
science made it easier for citizens to deceive themselves that climate change can
be treated as a non-issue, thus contributing to the policy impasse.

20. At the individual level, psychoanalysis is intended as a carefully calibrated
practice of therapeutic intervention, injecting reflexivity into patients and thus
helping them to confront their condition rather than living with the pathogenic
consequences of denial (see Friedrichs forthcoming, on pathological denial).

21. Alternative frames should have an empowering component, redefining the
problem in terms amenable to action. For example, people are more likely to
become active when something is framed as a health issue, as happened when
the ozone hole was defined as a health risk. Similarly, people accept the need for
action more easily when something is framed as an insurance problem. Invoking
the precautionary principle can activate their risk aversion. See the numerous
contributions about framing in the journal *Global Environmental Change*.

22. The true prophet knows that the task is not to convert the masses, but to
separate the sheep from the goats (Isaiah 6: 9–10).

23. The positive view of "vital lies" is shared by social psychologists (Goleman
1985; Taylor and Brown 1988; Jopling 1996; Lazarus 1998), and even philosophers sometimes acknowledge their benefits (Rorty 1994).

Chapter 7

1. The definition of transformability by Folke et al. (2011) captures the aspect
of resilience better than the classical definition by Walker et al. (2004, 1): "Trans-

formability is the capacity to create a fundamentally new system when ecological, economic, or social structures make the existing system untenable."

2. Giddens (1991) focuses on the individual self while others (e.g., Mitzen 2006) focus on state actors, but in the case of climate change and energy scarcity the focus must be on industrial civilization as a whole (Homer-Dixon 2006; for some conceptual groundwork, see Dalby 2009; Mayer and Schouten 2012).

3. In the 2012 US presidential campaign, both candidates chose to avoid the topic of climate change, thus enabling denial among citizens until Hurricane Sandy forced a public debate.

References

Adelman, M. A. 1995. *The Genie out of the Bottle: World Oil since 1970*. Cambridge, MA: MIT Press.

Adelman, M. A. 2004. The real oil problem. *Regulation* 27 (1):16–21.

Adger, W. Neil, Katrina Brown, and James Waters. 2011. Resilience. In *The Oxford Handbook of Climate Change and Society*, ed. J. S. Dryzek, K. M. Norgaard, and D. Schlosberg, 696–710. Oxford: Oxford University Press.

Agamben, Giorgio. 2005. *State of Exception*. Chicago: University of Chicago Press.

Agrawala, Shardul. 1998. Structural and process history of the Intergovernmental Panel on Climate Change. *Climatic Change* 39 (4):621–642.

Aimers, James, and David Hodell. 2011. Drought and the Maya. *Nature* 479:44–45.

Ainslie, George. 2001. *Breakdown of Will*. Cambridge: Cambridge University Press.

Akerlof, George A. 1991. Procrastination and obedience. *American Economic Review* 81 (2):1–19.

Alayarian, Aida. 2008. *Consequences of Denial: The Armenian Genocide*. London: Karnac.

Aleklett, Kjell, Mikael Höök, Kristofer Jakobsson, Michael Lardelli, Simon Snowden, and Bengt Söderbergh. 2010. The peak of the oil age: Analyzing the world oil production Reference Scenario in *World Energy Outlook 2008*. *Energy Policy* 38 (3):1398–1414.

Allen, Robert C. 2012. Backward into the future: The shift to coal and implications for the next energy transition. *Energy Policy* 50: 17–23.

Altieri, Miguel A., Nelso Companioni, Kristina Cañizares, Catherine Murphy, Peter Rosset, Martin Bourque, and Clara I. Nicholls. 1999. The greening of the "barrios": Urban agriculture for food security in Cuba. *Agriculture and Human Values* 16 (2):131–140.

Alvarez, José. 2004. *Cuba's Agricultural Sector*. Gainesville: University Press of Florida.

Anderegg, William R. L., James W. Prall, Jacob Harold, and Stephen H. Schneider. 2010. Expert credibility in climate change. *Proceedings of the National Academy of Sciences of the United States of America* 107 (27):12107–12109.

Anderson, David G., Kirk A. Maasch, and Daniel H. Sandweiss, eds. 2007. *Climate Change and Cultural Dynamics: A Global Perspective on Mid-Holocene Transitions.* London: Elsevier.

Archer, David, and Stefan Rahmstorf. 2010. *The Climate Crisis: An Introductory Guide to Climate Change.* Cambridge: Cambridge University Press.

Ayres, Robert U., and Benjamin Warr. 2009. *The Economic Growth Engine: How Energy and Work Drive Material Prosperity.* Cheltenham, UK: Edward Elgar.

Bächler, Günther, Voker Böge, Stefan Klötzli, Stefan Libiszewski, and Kurt R. Spillmann, eds. 1996. *Kriegsursache Umweltzerstörung.* 3 vols. Zürich: Rüegger.

Bäckstrand, Karin. 2003. Civic science for sustainability: Reframing the role of experts, policy-makers and citizens in environmental governance. *Global Environmental Politics* 3 (4):24–41.

Badal, Lionel. 2010. How the global oil watchdog failed its mission. *Countercurrents.org*, May 25. http://www.countercurrents.org/badal250510.htm.

Bamberger, Craig S. 2004. *The History of the International Energy Agency: The First Thirty Years.* Paris: OECD.

Bardi, Ugo. 2011. *The Limits to Growth Revisited.* New York: Springer.

Barnett, Jon, and W. Neil Adger. 2007. Climate change, human security and violent conflict. *Political Geography* 26 (6):639–655.

Barnhart, Michael A. 1987. *Japan Prepares for Total War: The Search for Economic Security, 1919–1941.* Ithaca, NY: Cornell University Press.

Barnhart, Michael A. 1995. *Japan and the World since 1868.* London: Edward Arnold.

Barry, John. 2012. *The Politics of Actually Existing Unsustainability: Human Flourishing in a Climate-Changed, Carbon-Constrained World.* Oxford: Oxford University Press.

Beasley, William G. 1987. *Japanese Imperialism, 1894–1945.* Oxford: Clarendon.

Beck, Ulrich. 1992. *Risk Society: Towards a New Modernity.* London: Sage.

Beck, Ulrich. 1999. *World Risk Society.* Cambridge: Polity.

Beck, Ulrich. 2009. *World at Risk.* Cambridge: Polity.

Behringer, Wolfgang. 2010. *A Cultural History of Climate.* Cambridge: Polity.

Benes, Jaromir, Marcelle Chauvet, Ondra Kamenik, Michael Kumhof, Douglas Laxton, Susanna Mursula, and Jack Selody. 2012. The Future of Oil: Geology versus Technology. IMF Working Paper 12/109.

Bentley, Roger. 2011. Roger Bentley. In *Peak Oil Personalities,* ed. C. J. Campbell, 31–52. Skibbereen, Ireland: Inspire.

Berkes, Fikret, Johan Colding, and Carl Folke, eds. 2003. *Navigating Social-Ecological Systems: Building Resilience for Complexity and Change*. Cambridge: Cambridge University Press.

Bernauer, Thomas, Tobias Böhmelt, and Vally Koubi. 2012. Environmental changes and violent conflict. *Environmental Research Letters* 7 (1).

Bierman, F., K. Abbott, S. Andresen, K. Bäckstrand, S. Bernstein, M. M. Betsill, H. Bulkeley, et al. 2012. Navigating the anthropocene: Improving earth system governance. *Science* 335:1306–1307.

Binningsbø, Helga Malmin, Indra de Soysa, and Nils Petter Gleditsch. 2007. Green giant or straw man? Environmental pressure and civil conflict, 1961–99. *Population and Environment* 28 (6):337–353.

Blühdorn, Ingolfur. 2010. The Politics of Unsustainability: Copenhagen, Post-Ecologism and the Performance of Seriousness. Paper presented at the ISA 51st Annual Convention, New Orleans, February 17–20, 2010.

Blühdorn, Ingolfur. 2011. The politics of unsustainability: COP15, post-ecologism, and the ecological paradox. *Organization & Environment* 24 (1):34–53.

Bolin, Bert. 2007. *A History of the Science and Politics of Climate Change: The Role of the Intergovernmental Panel on Climate Change*. Cambridge: Cambridge University Press.

Bradford, Travis. 2006. *Solar Revolution: The Economic Transformation of the Global Energy Industry*. Cambridge, MA: MIT Press.

Bradshaw, Michael J. 2010. Global energy dilemmas: a geographical perspective. *Geographical Journal* 176 (4):275–290.

Brandt, Adam R. 2007. Testing Hubbert. *Energy Policy* 35 (5):3074–3088.

Brandt, Adam R. 2012. Review of mathematical models of future oil supply: Historical overview and synthesizing critique. *Energy Bulletin* 35 (9): 3958–3974.

Brandt, Adam R., and Alexander E. Farrell. 2007. Scraping the bottom of the barrel: Greenhouse gas emission consequences of a transition to low-quality and synthetic petroleum resources. *Climatic Change* 84 (3):241–263.

Bray, Dennis, and Hans von Storch. 1999. Climate science: an empirical example of postnormal science. *Bulletin of the American Meteorological Society* 80 (3):439–455.

Brooks, Nick. 2006. Cultural responses to aridity in the middle Holocene and increased social complexity. *Quaternary International* 151 (1):29–49.

Brooks, Nick. 2010. Human responses to climatically-driven landscape change and resource scarcity: learning from the past and planning for the future. In *Landscapes and Societies: Selected Cases*, ed. I. P. Martini and W. Chesworth, 43–66. Dordrecht: Springer.

Brown, Lester Russell. 2011. *World on the Edge: How to Prevent Environmental and Economic Collapse*. London: Earthscan.

Brown, Marilyn A., and Benjamin K. Sovacool. 2011. *Climate Change and Global Energy Security: Technology and Policy Options.* Cambridge, MA: MIT Press.

Bryner, Gary. 2012. *Integrating Climate, Energy, and Air Pollution Policies.* Cambridge, MA: MIT Press.

Buckley, Brendan M., Kevin J. Anchukaitis, Daniel Penny, Roland Fletcher, Edward R. Cook, Masaki Sano, Le Canh Nam, et al. 2010. Climate as a contributing factor in the demise of Angkor, Cambodia. *Proceedings of the National Academy of Sciences of the United States of America* 107 (15):6748–6752.

Buhaug, Halvard. 2010. Climate not to blame for African civil wars. *Proceedings of the National Academy of Sciences of the United States of America* 107 (38):16477–16482.

Buhaug, Halvard, Nils Petter Gleditsch, and Ole Magnus Theisen. 2010. Implications of climate change for armed conflict. In *Social Dimensions of Climate Change: Equity and Vulnerability in a Warming World*, ed. R. Mearns and A. Norton, 75–101. Washington, DC: World Bank.

Burchardt, Hans-Jürgen, ed. 2000. *La última reforma agraria del siglo: La agricultura cubana entre el cambio y el estancamiento.* Caracas: Nueva Sociedad.

Burke, Marshall B., Edward Miguel, Shanker Satyanath, John A. Dykema, and David B. Lobell. 2009. Warming increases the risk of civil war in Africa. *Proceedings of the National Academy of Sciences of the United States of America* 106 (49):20670–20674.

Busby, Joshua W., Todd B. Smith, Kaiba L. White, and Shawn M. Strange. 2012. Locating climate insecurity: Where are the most vulnerable places in Africa? In *Climate Change, Human Security and Violent Conflict: Challenges for Societal Stability*, ed. J. Scheffran, M. Brzoska, H. G. Brauch, P. M. Link, and J. Schilling, 463–511. Heidelberg: Springer.

Buzan, Barry, Ole Wæver, and Jaap de Wilde. 1998. *Security: A New Framework for Analysis.* Boulder, CO: Lynne Rienner.

Campbell, Colin J. 1991. *The Golden Century of Oil 1950–2050: The Depletion of a Resource.* Dordrecht: Kluwer.

Campbell, Colin J. 1997. *The Coming Oil Crisis.* Brentwood, CA: Multiscience Publishing Company and Petroconsultants S.A.

Campbell, Colin J. 2011. Colin Campbell. In *Peak Oil Personalities*, ed. C. J. Campbell. Skribbereen, Ireland: Inspire.

Campbell, Colin J., and Jean H. Laherrère. 1998. The end of cheap oil: Global production of conventional oil will begin to decline sooner than most people think, probably within 10 years. *Scientific American* 278 (3):78–83.

Campbell, Kurt M., ed. 2008. *Climatic Cataclysm: The Foreign Policy and National Security Implications of Climate Change.* Washington, DC: Brookings Institution Press.

Canan, Penelope, and Nancy Reichman. 2002. *Ozone Connections: Expert Networks in Global Environmental Governance.* Sheffield: Greenleaf.

Carbon Tracker Initiative. 2011. Unburnable Carbon: Are the World's Financial Markets Carrying a Carbon Bubble? London: Carbon Tracker.

Carrasco, Alejandrina, David Acker, and James Grieshop. 2003. Absorbing the shocks: The case of food security, extension and the agricultural knowledge and information system in Havana, Cuba. *Journal of Agricultural Education and Extension* 9 (3):93–102.

Catton, William R. 1980. *Overshoot: The Ecological Basis of Revolutionary Change*. Urbana: University of Illinois Press.

Chaisson, Eric J. 2001. *Cosmic Evolution: The Rise of Complexity in Nature*. Cambridge, MA: Harvard University Press.

Chang, Yu Sang, and Seung Jin Baek. 2010. Limit to improvement: Myth or reality? Empirical analysis of historical improvement on three technologies influential in the evolution of civilization. *Technological Forecasting and Social Change* 77 (5):712–729.

Chapin III, F. Stuart, Stephen R. Carpenter, Gary P. Kofinas, Carl Folke, Nick Abel, William C. Clark, Per Olsson, et al. 2009. Ecosystem stewardship: sustainability strategies for a rapidly changing planet. *Trends in Ecology & Evolution* 25 (4):241–249.

Chew, Sing C. 2007. *The Recurring Dark Ages: Ecological Stress, Climate Changes, and System Transformation*. Lanham, MD: AltaMira.

Chew, Sing C. 2008. *Ecological Futures: What History Can Teach Us*. Lanham, MD: AltaMira.

Chisholm, Don. 1995. A troubleshooter's analysis of the human predicament. *Futures* 27 (3):353–362.

Choucri, Nazli, Michael Laird, and Dennis L. Meadows. 1972. Resource Sacrcity and Foreign Policy: A Simulation Model of International Conflict. Cambridge, MA: MIT Center for International Studies.

Christian, David. 2004. *Maps of Time: An Introduction to Big History*. Berkeley, CA: University of California Press.

CIA. 1996. *Cuba: Handbook of Trade Statistics, 1996*. Washington, DC: Central Intelligence Agency.

Clarke, Duncan. 2007. *The Battle for Barrels: Peak Oil Myths and World Oil Futures*. London: Profile.

Cleveland, Cutler J. 2005. Net energy from the extraction of oil and gas in the United States. *Energy* 30 (5):769–782.

CNA Corporation. 2007. National Security and the Threat of Climate Change. Alexandria, VA: CNA Corporation.

Cobb, James C. 1984. *Industrialization and Southern Society, 1877–1984*. Lexington: University Press of Kentucky.

Cohen, Joel E. 1995. *How Many People Can the Earth Support*. New York: Norton.

Cohen, Stanley. 2001. *States of Denial: Knowing about Atrocities and Suffering*. Cambridge: Polity.

Colgan, Jeff. 2009. The International Energy Agency: Challenges for the 21st century. Policy Paper N. 6. Berlin: Global Public Policy Institute.

Collier, David, James Mahoney, and Jason Seawright. 2004. Claiming too much: Warnings about selection bias. In *Rethinking Social Inquiry: Diverse Tools, Shared Standards*, ed. H. E. Brady and D. Collier, 85–102. Lanham, MD: Rowman and Littlefield.

Conkling, Philip, Richard Alley, Wallace Broecker, and George Denton. 2011. *The Fate of Greenland: Lessons from Abrupt Climate Change.* Cambridge, MA: MIT Press.

Council on Environmental Quality and US Department of State. 1980. *The Global 2000 Report to the President: Entering the Twenty-First Century.* Washington, DC: US Government Printing Office.

Cruz, María Caridad, and Roberto Sánchez Medina. 2003. *Agriculture in the City: A Key to Sustainability in Havana, Cuba.* Kingston, Jamaica: Ian Randle.

Dalby, Simon. 2009. *Security and Environmental Change.* Cambridge: Polity.

Dale, Michael, Susan Krumdieck, and Pat Bodger. 2011. Net energy yield from production of conventional oil. *Energy Policy* 39 (11):7095–7102.

Dasgupta, Partha. 2008. Discounting climate change. *Journal of Risk and Uncertainty* 37 (2):141–169.

David, Paul A. 1985. Clio and the economics of QWERTY. *Economy and History* 75 (2):332–337.

de Almeida, Pedro, and Pedro D. Silva. 2011. Timing and future consequences of the peak of oil production. *Futures* 43 (10):1044–1055.

De Marchi, Bruna, and Jerome R. Ravetz. 1999. Risk management and governance: A post-normal science approach. *Futures* 31 (7):743–757.

De Young, Raymond, and Thomas Princen, eds. 2012. *The Localization Reader: Adapting to the Coming Downshift.* Cambridge, MA: MIT Press.

Death, Carl. 2011. Summit theatre: Exemplary governmentality and environmental diplomacy in Johannesburg and Copenhagen. *Environmental Politics* 20 (1):1–19.

Deere, Carmen Diana, Niurka Pérez, and Ernel González. 1994. The view from below: Cuban agriculture in the "Special Period in Peacetime." *Journal of Peasant Studies* 21 (2):194–234.

Deffeyes, Kenneth S. 2001. *Hubbert's Peak: The Impending World Oil Shortage.* Princeton, NJ: Princeton University Press.

Delucchi, Mark A., and Mark Z. Jacobson. 2011. Providing all global energy with wind, water, and solar power, Part II: Reliability, system and transmission costs, and policies. *Energy Policy* 39 (3):1170–1190.

deMenocal, Peter B. 2001. Cultural responses to climate change during the late Holocene. *Science* 292:667–673.

Dessler, Andrew E. 2012. *Introduction to Modern Climate Change.* Cambridge: Cambridge University Press.

Detraz, Nicole, and Michele M. Betsill. 2009. Climate change and environmental security: For whom the discourse shifts. *International Studies Perspectives* 10 (3):303–320.

Diamond, Jared. 2005. *Collapse: How Societies Choose to Fail or Survive.* London: Allen Lane.

Diamond, Jared. 2009. Maya, Khmer and Inca. *Nature* 461:479–480.

Díaz Briquets, Sergio, and Jorge Pérez-López. 2000. *Conquering Nature: The Environmental Legacy of Socialism in Cuba.* Pittsburgh, PA: University of Pittsburgh Press.

Dilworth, Craig. 2010. *Too Smart for Our Own Good: The Ecological Predicament of Humankind.* Cambridge: Cambridge University Press.

DiMuzio, Tim. 2012. Capitalizing a future unsustainable: Finance, energy and the fate of market civilization. *Review of International Political Economy* 19 (3):363–388.

Dinar, Shlomi, ed. 2011. *Beyond Resource Wars: Scarcity, Environmental Degradation, and International Cooperation.* Cambridge, MA: MIT Press.

Dosi, Giovanni, and Marco Grazzi. 2009. Energy, development and the environment: An appraisal three decades after the "Limits to growth" debate. In *Recent Advances in Neo-Schumpeterian Economics: Essays in Honour of Horst Hanusch,* ed. A. Pyka, U. Cantner, A. Greiner, and T. Kuhn, 34–52. Cheltenham, UK: Edward Elgar.

Dryzek, John S. 2005. *The Politics of the Earth: Environmental Discourses.* 2nd ed. Oxford: Oxford University Press.

Dunlap, Riley E., and Aaron M. McCright. 2011. Organized climate change denial. In *Oxford Handbook of Climate Change and Society,* ed. J. S. Dryzek, R. B. Norgaard, and D. Schlosberg, 144–160. Oxford: Oxford University Press.

Dunning, Thad. 2008. *Crude Democracy: Natural Resource Wealth and Political Regimes.* Cambridge: Cambridge University Press.

Dyer, Gwynne. 2010. *Climate Wars: The Fight for Survival as the World Overheats.* 2nd ed. Oxford: Oneworld.

Dyke, Daniel. 1614. *The Mystery of Selfe-deceiving: A Discourse and Discouerie of the Deceitfullnesse of Mans Heart.* London: Edward Griffin.

Eastin, Josh, Reiner Grundmann, and Aseem Prakash. 2011. The two limits debates: "Limits to Growth" and climate change. *Futures* 43 (1):16–26.

Eberstadt, Nicholas. 1999. *The End of North Korea.* Washington, DC: AEI Press.

Eberstadt, Nicholas. 2009. *The North Korean Economy: Between Crisis and Catastrophe.* New Brunswick, NJ: Transaction.

Economist Intelligence Unit. 2008. *Country Profile: Cuba.* London: Economist Intelligence Unit.

Ehrlich, Paul R., and Anne H. Ehrlich. 1998. *Betrayal of Science and Reason: How Anti-Environmental Rhetoric Threatens Our Future.* Washington, DC: Island Press.

Ehrlich, Paul R., and Anne H. Ehrlich. 2004. *One with Nineveh: Politics, Consumption, and the Human Nature*. Washington, DC: Island Press.

Ehrlich, Paul R., Gary Wolff, Gretchen C. Daily, Jennifer B. Hughes, Scott Daily, Michael Dalton, and Lawrence Goulder. 1999. Knowledge and the environment. *Ecological Economics* 30 (2):267–284.

Elhefnawy, Nader. 2008. The impending oil shock. *Survival* 50 (2):37–66.

Energy Watch Group. 2007. Coal: Resources and future production. Ottobrunn, Germany: Energy Watch Group.

Ewing, Brad, David Moore, Steven Goldfinger, Anna Oursler, Anders Reed, and Mathis Wackernagel. 2010. *The Ecological Footprint Atlas 2010*. Oakland, CA.: Global Footprint Network.

Fagan, Brian. 2000. *The Little Ice Age: How Climate Made History, 1300–1850*. New York: Basic Books.

Fagan, Brian. 2004. *The Long Summer: How Climate Changed Civilization*. London: Granta.

Fagan, Brian. 2008. *The Great Warming: Climate Change and the Rise and Fall of Civilizations*. New York: Bloomsbury.

Fagan, Brian. 2009. *Floods, Famines and Emperors: El Niño and the Fate of Civilizations*. London: Pimlico.

Fagan, Mary. 2000. Sheikh Yamani predicts price crash as age of oil ends. *Sunday Telegraph*, June 25, 2000.

Fantazzini, Dean, Mikael Höök, and André Angelantoni. 2011. Global oil risks in the early 21st century. *Energy Policy* 39 (12):7865–7873.

FAO. 2003. Fertilizer Use by Crop in Cuba. Rome: Food and Agriculture Organization.

FAO/WFP. 1999. Special Report: FAO/WFP Crop and Food Supply Assessment Mission to the Democratic People's Republic of Korea, November 8, 1999.

FAO/WFP. 2008. Special Report: FAO/WFP Crop and Food Security Assessment Mission to the Democratic People's Republic of Korea, December 8, 2008.

Fitzgerald, Michael W. 2007. *Splendid Failure: Postwar Reconstruction in the American South*. Chicago: Ivan R. Dee.

Fogel, Robert William. 1989. *Without Consent or Contract: The Rise and Fall of American Slavery*. New York: Norton.

Folke, Carl, Stephen R. Carpenter, Brian Walker, Marten Scheffer, Terry Chapin, and Johan Rockström. 2010. Resilience thinking: Integrating resilience, adaptability and transformability. *Ecology and Society* 15 (4).

Folke, Carl, Åsa Jansson, Johan Rockström, Per Olsson, Stephen R. Carpenter, F. Stuart Chapin III, Anne-Sophie Crépin, et al. 2011. Reconnecting to the biosphere. *Ambio* 40 (7):719–738.

Fouquet, Roger, and Peter J. G. Pearson. 2012. Past and prospective energy transitions: Insights from history. *Energy Policy* 50: 1–7.

Friedman, Benjamin M. 2005. *The Moral Consequences of Economic Growth*. New York: Alfred A. Knopf.

Friedrichs, Jörg. 2010. Global energy crunch: How different parts of the world would react to a peak oil scenario. *Energy Policy* 38 (8):4562–4569.

Friedrichs, Jörg. 2011. Peak energy and climate change: The double bind of postnormal science. *Futures* 43 (4):469–477.

Friedrichs, Jörg. Forthcoming. Useful lies: The twisted rationality of denial. *Philosophical Psychology* 26.

Friedrichs, Jörg. Forthcoming. Who's afraid of Thomas Malthus? In *Understanding Society and Natural Resources: Forging New Strands of Integration Across the Social Sciences*, ed. M. J. Manfredo, A. Rechkemmer, and J. J. Vaske. New York: Springer.

Funes, Fernando, Luis García, Martin Bourque, Nilda Pérez, and Peter Rosset, eds. 2002. *Sustainable Agriculture and Resistance: Transforming Food Production in Cuba*. Oakland, CA: First Books.

Funtowicz, Silvio O., and Jerome R. Ravetz. 1993. Science for the post-normal age. *Futures* 25 (7):739–755.

Gagnon, Nathan, Charles A.S. Hall, and Lysle Brinker. 2009. A preliminary investigation of Energy Return on Energy Investment for global oil and gas production. *Energies* 2 (3):490–503.

Gautier, Catherine. 2008. *Oil, Water, and Climate: An Introduction*. Cambridge: Cambridge University Press.

Giddens, Anthony. 1990. *The Consequences of Modernity*. Cambridge: Polity.

Giddens, Anthony. 1991. *Modernity and Self-Identity: Self and Society in the Late Modern Age*. Cambridge: Polity.

Giddens, Anthony. 2009. *The Politics of Climate Change*. Cambridge: Polity.

Gilpin, Robert. 1987. *The Political Economy of International Relations*. Princeton, NJ: Princeton University Press.

Gleditsch, Nils Petter. 2012. Wither the weather? Climate change and conflict. *Journal of Peace Research* 49 (1):3–9.

Godfray, H. Charles J., John R. Beddington, Ian R. Crute, Lawrence Haddad, David Lawrence, James F. Muir, Jules Pretty, et al. 2010. Food security: The challenge of feeding 9 billion people. *Science* 327:812–818.

Goleman, Daniel. 1985. *Vital Lies, Simple Truths: The Psychology of Self-Deception*. New York: Simon and Schuster.

Goodkind, Daniel, and Daniel West. 2001. The North Korean famine and its demographic impact. *Population and Development Review* 27 (2):219–238.

Gore, Al. 2006. *An Inconvenient Truth: The Planetary Emergency of Global Warming and What We Can Do about It*. London: Bloomsbury.

Gorelick, Steven M. 2010. *Oil Panic and the Global Crisis: Predictions and Myths*. Chichester, UK: Wiley-Blackwell.

Gort, Michael, and Richard A. Wall. 1986. The evolution of technologies and investment in innovation. *Economic Journal* 96 (383):741–757.

Greenpeace. 2010. *Koch Industries Secretly Funding the Climate Denial Machine.* Washington, DC: Greenpeace USA.

Greenpeace. 2012. EXXONSECRETS.ORG: How ExxonMobil funds the climate change skeptics. http://www.exxonsecrets.org.

Greer, John Michael. 2008. *The Long Descent: A User's Guide to the End of the Industrial Age.* Gabriola Island, BC. New Society Publishers.

Greer, John Michael. 2009. *The Ecotechnic Future: Envisioning a Post-Peak World.* Gabriola Island, BC. New Society Publishers.

Grundmann, Reiner. 2007. Climate change and knowledge politics. *Environmental Politics* 16 (3):414–432.

Guilford, Megan C., Charles A. S. Hall, Pete O'Connor, and Cutler J. Cleveland. 2011. A new long term assessment of energy return on investment (EROI) for U.S. oil and gas discovery and production. *Sustainability* 3 (10):1866–1887.

Gupta, Ajay K., and Charles A. S. Hall. 2011. A review of the past and current state of EROI data. *Sustainability* 3 (10):1796–1809.

Gurr, Ted Robert. 1985. On the political consequences of scarcity and economic decline. *International Studies Quarterly* 29 (1):51–75.

Haberl, Helmut, Marina Fischer-Kowalski, Fridolin Krausmann, Joan Martinez-Alier, and Verena Winiwarter. 2011. A socio-metabolic transition towards sustainability? Challenges for another great transformation. *Sustainable Development* 19 (1):1–14.

Haggard, Stephen, and Marcus Noland. 2007. *Famine in North Korea: Markets, Aid, and Reform.* New York: Columbia University Press.

Hall, Charles A. S., and John W. Day. 2009. Revisiting the limits to growth after peak oil. *American Scientist* 97 (3):230–237.

Hamilton, Clive. 2010. *Requiem for a Species: Why We Resist the Truth about Climate Change.* London: Earthscan.

Hamilton, James D. 2009. Causes and consequences of the oil shock of 2007–08. *Brookings Papers on Economic Activity* (Spring):215–261.

Hamilton, James D. 2011. Historical Oil Shocks. National Bureau of Economic Research. Working Paper 16790.

Hamilton, James D. 2013. Historical oil shocks. In *The Routledge Handbook of Major Events in Economic History*, ed. R. E. Parker and R. Whaples, 239–265. London and New York: Routledge.

Hansen, James E. 2009. *Storms of My Grandchildren: The Truth about the Coming Climate Catastrophe and Our Last Chance to Save Humanity.* London: Bloomsbury.

Hardin, Garrett. 1968. The Tragedy of the Commons. *Science* 162:1243–1248.

Hardin, Russell. 1982. *Collective Action.* Baltimore, MD: Johns Hopkins University Press.

Hauge, Wenche, and Tanja Ellingsen. 1998. Beyond environmental scarcity: Causal pathways to conflict. *Journal of Peace Research* 35 (3):299–317.

Havel, Vaclav. 1989. *Living in Truth*. London: Faber.

Healy, Stephen. 1999. Extended peer communities and the ascendance of post-normal politics. *Futures* 31 (7):655–669.

Heinberg, Richard. 2003. *The Party's Over: Oil, War, and the Fate of Industrial Societies*. Gabriola Island, BC: New Society Publishers.

Heinberg, Richard. 2004. *Powerdown: Options and Actions for a Post-Carbon World*. Gabriola Island, BC: New Society Publishers.

Heinberg, Richard. 2006. *The Oil Depletion Protocol: A Plan to Avert Oil Wars, Terrorism and Economic Collapse*. Gabriola Island, BC: New Society Publishers.

Heinberg, Richard. 2007. *Peak Everything: Waking up to the Century of Decline in Earth's Resources*. Gabriola Island, BC: New Society Publishers.

Heinberg, Richard. 2009a. *Blackout: Coal, Climate and the Last Energy Crisis*. Gabriola Island, BC: New Society Publishers.

Heinberg, Richard. 2009b. *Searching for a Miracle: "Net Energy" Limits and the Fate of Industrial Society*. San Francisco: International Forum on Globalization and Post-Carbon Institute.

Heinberg, Richard. 2011. *The End of Growth: Adapting to Our New Economic Reality*. Gabriola Island, BC. New Society Publishers.

Heinberg, Richard, and Daniel Lerch, eds. 2010. *The Post Carbon Reader: Managing the 21st Century's Sustainability Crisis*. Healdsburg, CA.: Watershed Media.

Helm, Dieter. 2012. *The Carbon Crunch: How We're Getting Climate Change Wrong, and How to Fix It*. New Haven, CT: Yale University Press.

Hendrix, Cullen S., and Idean Salehyan. 2012. Climate change, rainfall, and social conflict in Africa. *Journal of Peace Research* 49 (1):35–50.

Herman Jr., Paul F., and Gregory F. Treverton. 2009. The political consequences of climate change. *Survival* 51 (2):137–148.

Heun, Matthew Kuperus, and Martin de Wit. 2012. Energy return on (energy) invested (EROI), oil prices, and energy transitions. *Energy Policy* 40 (1):147–158.

Hirsch, Robert L. 2008. Mitigation of maximum world oil production: Shortage scenarios. *Energy Policy* 36 (2):881–889.

Hirsch, Robert L., Roger Bezdek, and Robert Wendling. 2005. *Peaking of World Oil Production: Impacts, Mitigation and Risk Management*. http://www.netl.doe.gov/publications/others/pdf/Oil_Peaking_NETL.pdf.

Hirsch, Robert L., Roger Bezdek, and Robert Wendling. 2010. *The Impending World Energy Mess*. Burlington, Ontario: Griffin Media.

Hoffman, Andrew J. 2011. Talking past each other? Cultural framing of skeptical and convinced logics in the climate change debate. *Organization & Environment* 24 (1):3–33.

Hoffmann, Peter. 2012. *Tomorrow's Energy: Hydrogen, Fuel Cells, and the Prospects for a Cleaner Planet.* Cambridge, MA: MIT Press.

Hoggan, James. 2009. *Climate Cover-Up: The Crusade to Deny Global Warming.* Vancouver, BC: Greystone.

Holmgren, David. 2009. *Future Scenarios: How Communities Can Adapt to Peak Oil and Climate Change.* Devon, UK: Green Books.

Homer-Dixon, Thomas F. 1994. Environmental scarcities and violent conflict: evidence from cases. *International Security* 19 (1):5–40.

Homer-Dixon, Thomas F. 1995. The ingenuity gap: Can poor countries adapt to resource scarcity? *Population and Development Review* 21 (3):587–612.

Homer-Dixon, Thomas F. 1999. *Environment, Scarcity, and Violence.* Princeton, NJ: Princeton University Press.

Homer-Dixon, Thomas F. 2001. *The Ingenuity Gap.* London: Vintage.

Homer-Dixon, Thomas F. 2006. *The Upside of Down: Catastrophe, Creativity and the Renewal of Civilization.* London: Souvenir Press.

Homer-Dixon, Thomas F., ed. 2009. *Carbon Shift: How the Twin Crises of Oil Depletion and Climate Change Will Define the Future.* Toronto: Random House Canada.

Höök, Mikael, and Kjell Aleklett. 2009. Historical trends in American coal production and a possible future outlook. *International Journal of Coal Geology* 78 (3):201–216.

Höök, Mikael, Juchen Li, Kersti Johansson, and Simon Snowden. 2012. Growth rates of global energy systems and future outlooks. *Natural Resources Research* 21 (1):23–41.

Höök, Mikael, and Xu Tang. 2013. Depletion of fossil fuels and anthropogenic climate change: A review. *Energy Policy* 52:797–809.

Höök, Mikael, Werner Zittel, Jörg Schindler, and Kjell Aleklett. 2010. Global coal production outlooks based on a logistic model. *Fuel* 89 (11):3546–3558.

Hopkins, Rob. 2008. *The Transition Handbook: From Oil Dependency to Local Resilience.* Totnes, UK: Green Books.

Houghton, John. 2009. *Global Warming: The Complete Briefing.* 4th ed. Cambridge: Cambridge University Press.

Hsiang, Solomon M., Kyle C. Meng, and Mark A. Cane. 2011. Civil conflicts are associated with the global climate. *Nature* 476:438–441.

Hubbert, M. King. 1956. Nuclear energy and the fossil fuels. Paper presented to the American Petroleum Institute, March 8, 1956.

Hubbert, M. 1962. *Energy Resources: A Report to the Committee on Natural Resources.* Washington, DC: National Academy of Sciences.

Hubbert, M. 1981. The world's evolving energy system. *American Journal of Physics* 49 (11):1007–1029.

Hubbert, M. 1993. Exponential growth as a transient phenomenon in human history. In *Valuing the Earth: Economics, Ecology, Ethics*, ed. H. E. Daly and K. N. Townsend, 113–126. Cambridge, MA: MIT Press.

Hulme, Mike. 2009. *Why We Disagree about Climate Change: Understanding Controversy, Inaction and Opportunity*. Cambridge: Cambridge University Press.

Hume, David. 1739. *A Treatise of Human Nature*, vol. 2. 3 vols. London: John Noon.

Hume, David. [1748] 2000. *An Enquiry Concerning Human Understanding*, ed. Tom L. Beauchamp. Oxford: Clarendon.

Ibsen, Henrik. [1884] 2006. *The Wild Duck*, trans. S. Mulrine. London: Nick Hern.

IEA. 1974. Agreement on an International Energy Program (as amended September 25, 2008).

IEA. 1998. *World Energy Outlook 1998*. Paris: International Energy Agency.

IEA. 2008. *World Energy Outlook 2008*. Paris: International Energy Agency.

IEA. 2009. *World Energy Outlook 2009*. Paris: International Energy Agency.

IEA. 2010a. *World Energy Outlook 2010*. Paris: International Energy Agency.

IEA. 2010b. *Energy Technology Perspectives 2010: Scenarios and Strategies to 2050*. 2 vols. Paris: International Energy Agency.

IEA. 2011a. *World Energy Outlook 2011*. Paris: International Energy Agency.

IEA. 2011b. *CO2 Emissions from Fuel Combustion: Highlights*. Paris: International Energy Agency.

IEA. 2012a. *World Energy Outlook 2012*. Paris: International Energy Agency.

IEA. 2012b. *Energy Technology Perspectives 2012: Pathways to a Clean Energy System*. Paris: International Energy Agency.

IEA. 2012c. *Golden Rules for a Golden Age of Gas: World Energy Outlook Special Report on Unconventional Gas*. Paris: International Energy Agency.

IEA. 2012d. *IEA Response System for Oil Supply Emergencies*. Paris: International Energy Agency.

IEA. 2012e. *Key World Energy Statistics*. Paris: International Energy Agency.

IEA. 2012f. Employment at the IEA, July 7. http://www.iea.org/about/job.asp.

Ike, Nobutaka. 1967. *Japan's Decision for War: Records of the 1941 Policy Conferences*. Stanford, CA: Stanford University Press.

IMF. 2011. *Tensions from the Two-Speed Recovery: Unemployment, Commodities, and Capital Flows.* World Economic Outlook (April). Washington, DC: International Monetary Fund.

IPCC. 1990. *Climate Change: The IPCC Scientific Assessment*. Cambridge: Cambridge University Press.

IPCC. 1995. *Climate Change 1995: IPCC Second Assessment*. Geneva: Intergovernmental Panel on Climate Change.

IPCC. 2001a. *Climate Change 2001: Synthesis Report.* Cambridge: Cambridge University Press.

IPCC. 2001b. *Climate Change 2001: The Scientific Basis.* Cambridge: Cambridge University Press.

IPCC. 2001c. *Climate Change 2001: Impacts, Adaptation and Vulnerability. Working Group II Contribution to the Third Assessment Report of the Intergovernmental Panel on Climate Change.* Cambridge: Cambridge University Press.

IPCC. 2007a. *Climate Change 2007: Synthesis Report.* Geneva: Intergovernmental Panel on Climate Change.

IPCC. 2007b. *Climate Change 2007: Impacts, Adaptation and Vulnerability. Working Group II Contribution to the Fourth Assessment Report of the Intergovernmental Panel on Climate Change.* Cambridge: Cambridge University Press.

IPCC. 2012. *Managing the Risks of Extreme Weather Events and Disasters to Advance Climate Change Adaptation.* Cambridge: Cambridge University Press.

Jackson, Tim. 2009. *Prosperity without Growth: Economics for a Finite Planet.* London: Earthscan.

Jacobson, Mark Z., and Mark A. Delucchi. 2011. Providing all global energy with wind, water, and solar power, Part I: technologies, energy resources, quantities and areas of infrastructure, and materials. *Energy Policy* 39 (3): 1154–1169.

Jatar-Hausmann, Ana Julia. 1999. *The Cuban Way: Capitalism, Communism and Confrontation.* West Hartford, CT: Kumarian Press.

Jones, Gwyn. 1986. *The Norse Atlantic Saga: Being the Norse Voyages of Discovery and Settlement to Iceland, Greenland, and North America.* 2nd ed. Oxford: Oxford University Press.

Jopling, David A. 1996. "Take away the life-lie": Positive illusions and creative self-deception. *Philosophical Psychology* 9 (4):525–544.

Kahl, Colin H. 2006. *States, Scarcity, and Civil Strive in the Developing World.* Princeton, NJ: Princeton University Press.

Kamal, Sajed. 2011. *The Renewable Revolution: How we can fight climate change, prevent energy wars, revitalize the economy and transition to a sustainable future.* London: Earthscan.

Karlsson, Gunnar. 2000. *Iceland's 1100 Years: The History of a Marginal Society.* London: Hurst.

Kauffman, Stuart A. 1995. *At Home in the Universe: The Search for the Laws of Self-Organization and Complexity.* New York: Oxford University Press.

Kennett, Douglas J., and James P. Kennett. 2007. Influence of Holocene marine transgression and climate change on cultural evolution in southern Mesopotamia. In *Climate Change and Cultural Dynamics: A Global Perspective on Mid-Holocene Transitions*, ed. D. G. Anderson, K. A. Maasch, and D. H. Sandweiss, 229–264. London: Elsevier.

Keohane, Robert O. 1984. *After Hegemony: Cooperation and Discord in the World Political Economy.* Princeton, NJ: Princeton University Press.

Kerr, Richard A. 2012. Are world oil's prospects not declining all that fast? *Science* 337: 633.

Kershaw, Ian. 2007. *Fateful Choices: Ten Decisions that Changed the World*. New York: Penguin.

Keynes, John Maynard. 1923. *A Tract on Monetary Reform*. London: Macmillan.

Kindleberger, Charles P. 1973. *The World in Depression, 1929–1939*. Berkeley, CA: University of California Press.

Kissinger, Henry. 1973. Text of address by Kissinger in London on energy and European problems. *New York Times*, December 13.

Klare, Michael T. 2004. *Blood and Oil: The Dangers and Consequences of America's Growing Dependency on Imported Petroleum*. New York: Metropolitan Books.

Klare, Michael T. 2008. *Rising Powers, Shrinking Planet: The New Geopolitics of Energy*. New York: Metropolitan Books.

Klare, Michael T. 2012. *The Race for What's Left: The Global Scramble for the World's Last Resources*. New York: Metropolitan Books.

Kohl, Wilfrid L. 2010. Consumer country energy cooperation: The International Energy Agency and the global energy order. In *Global Energy Governance: The New Rules of the Game*, ed. A. Goldhau and J. M. Witte, 195–220. Washington, DC: Brookings Press.

Korowicz, David. 2010. Tipping Point: Near-Term Systemic Implications of a Peak in Global Oil Production. Feasta Working Paper. Dublin: Feasta.

Kuecker, Glen David, and Thomas D. Hall. 2011. Resilience and community in the age of world-system collapse. *Nature and Culture* 6 (1):18–40.

Kuhn, Thomas S. 1962. *The Structure of Scientific Revolutions*. Chicago: University of Chicago Press.

Kunstler, James Howard. 2005. *The Long Emergency: Surviving the Converging Catastrophes of the Twenty-First Century*. New York: Atlantic Monthly.

Kurzweil, Ray. 2006. *The Singularity Is Near: When Humans Transcend Biology*. London: Duckworth.

Lahsen, Myanna. 2008. Experiences of modernity in the greenhouse: A cultural analysis of a physicist "trio" supporting the backlash against global warming. *Global Environmental Change* 18 (1):204–219.

Lamb, Hubert H. 1977. *Climate: Present, Past, and Future. Climatic History and the Future*, vol. 2. London: Methuen.

Lang, Guenter. 2009. Measuring the returns of R&D: An empirical study of the German manufacturing sector over 45 years. *Research Policy* 38:1438–1445.

Lawler, Andrew. 2008. Indus collapse: The end or the beginning of an Asian culture? *Science* 320:1281–1283.

Lawler, Andrew. 2010. Collapse? What collapse? Societal change revisited. *Science* 330:907–909.

Lazarus, Richard S. 1998. The costs and benefits of denial. In *Fifty years of the Research and Theory of R.S. Lazarus: An Analysis of Historical and Perennial Issues*, 227–251. Mahwah, NJ: Erlbaum.

LeBlanc, Steven A. 2003. *Constant Battles: The Myth of the Peaceful, Noble Savage*. New York: St. Martin's Press.

Leder, Frederic, and Judith N. Shapiro. 2008. This time it's different: An inevitable decline in world petroleum production will keep oil product prices high, causing military conflicts and shifting wealth and power from democracies to authoritarian regimes. *Energy Policy* 36 (8):2850–2852.

Leggett, Jeremy. 2005. *Oil, Gas, Hot Air, and the Coming Global Financial Catastrophe*. New York: Random House.

Lenton, Timothy M., Hermann Held, Elmar Kriegler, Jim W. Hall, Wolfgang Lucht, Stefan Rahmstorf, and Hans Joachim Schellnhuber. 2008. Tipping elements in the Earth's climate system. *Proceedings of the National Academy of Sciences of the United States of America* 105 (6):1786–1793.

Levin, Simon. 2010. Crossing scales, crossing disciplines: Collective motion and collective action in the global commons. *Philosophical Transactions of the Royal Society of London. Series B, Biological Sciences* 365 (1):13–18.

Lewis, Oscar, Ruth M. Lewis, and Susan M. Rigdon. 1978. *Neighbors: Living the Revolution. An Oral History of Contemporary Cuba*. Urbana: University of Illinois Press.

Lin, Bo-qiang, and Jiang-hua Liu. 2010. Estimating coal production peak and trends of coal imports in China. *Energy Policy* 38 (1):512–519.

Lin, Boquiang, Jianhua Liu, and Yang Yinchun. 2012. Impact of carbon intensity and energy security constraints on China's coal import. *Energy Policy* 48: 137–147.

Lipschutz, Ronnie D. 1989. *When Nations Clash: Raw Materials, Ideology and Foreign Policy*. Cambridge, MA: Ballinger.

Lipson, Daniel N. 2011. Is the great recession only the beginning? Economic contraction in an age of fossil fuel depletion and ecological limits to growth. *New Political Science* 33 (4):555–575.

Litfin, Karen T. 1994. *Ozone Discourses: Science and Politics in Global Environmental Cooperation*. New York: Columbia University Press.

Lovelock, James. 2000. *The Ages of Gaia: A Biography of our Living Earth*. 2nd ed. Oxford: Oxford University Press.

Lovelock, James. 2006. *The Revenge of Gaia: Why the Earth is Fighting Back*. New York: Basic Books.

Lovelock, James. 2009. *The Vanishing Face of Gaia: A Final Warning*. London: Allen Lane.

Lowe, Thomas, Katrina Brown, Suraje Dessai, Doria Miguel de França, Kat Haynes, and Katharine Vincent. 2006. Does tomorrow ever come? Disaster narrative and public perceptions of climate change. *Public Understanding of Science* 15 (4):435–457.

Lutz, Christian, Ulrike Lehr, and Kirsten S. Wiebe. 2012. Economic effects of peak oil. *Energy Policy* 48: 829–834.

Lutz, Wolfgang, and Samir K.C. 2010. Dimensions of global population projections: What do we know about future population trends and structures? *Philosophical Transactions of the Royal Society of London. Series B, Biological Sciences* 365:2779–2791.

Lynas, Mark. 2007. *Six Degrees: Our Future on a Hotter Planet.* London: Harper Perennial.

Macalister, Terry. 2009. Whistleblower: Key oil figures were distorted by US pressure. *The Guardian,* November 10.

McCright, Aaron M., and Riley E. Dunlap. 2010. Anti-reflexivity: The American conservative movement's success in undermining climate science and policy. *Theory, Culture & Society* 27 (2):100–133.

McCright, Aaron M., and Riley E. Dunlap. 2011. The politicization of climate change and polarization in the American public's views of global warming, 2001–2010. *Sociological Quarterly* 52 (2):155–194.

McGaurr, Lyn. 2009. Putting the Globe in the Sphere: Climate Change Scientists in the Public Sphere. Paper presented to the 2009 ANZCA Conference.

McKenna, Laura. 2007. "Getting the word out": Policy bloggers use their soap box to make change. *Review of Policy Research* 24 (3):209–229.

MacKenzie, Debora. 2010. Living in denial: why sensible people reject the truth. *New Scientist* 206 (2760):38–41.

Malthus, Thomas. 1798. *An Essay on the Principle of Population, as it Affects the Future Improvement of Society.* London: J. Johnson.

Mann, Michael E. 2012. *The Hockey Stick and the Climate Wars: Dispatches from the Front Lines.* New York: Columbia University Press.

Mann, Michael E., Zhihua Zhang, Scott Rutherford, Raymond S. Bradley, Malcolm K. Hughes, Drew Shindell, Caspar Ammann, et al. 2009. Global signatures and dynamical origins of the Little Ice Age and medieval climate anomaly. *Science* 326:1256–1260.

Manning, Martin. 2011. The climate for science. In *The Governance of Climate Change: Science, Economics, Politics and Ethics,* ed. D. Held, A. Hervey, and M. Theros, 31–48. Cambridge: Polity.

Marsden, William. 2011. *Fools Rule: Inside the Failed Politics of Climate Change.* Toronto: Alfred A. Knopf Canada.

Maugeri, Leonardo. 2006. *The Age of Oil: The Mythology, History, and Future of the World's Most Controversial Resource.* Westport, CT: Praeger.

Maugeri, Leonardo. 2012. Oil: The Next Revolution. Belfer Center for Science and International Affairs, Cambridge, MA. Discussion Paper 2012-10.

Mayer, Maximilian, and Peer Schouten. 2012. Energy security and climate security under conditions of the Anthropocene. In *Energy Security in the Era of Climate Change: The Asia-Pacific Experience,* ed. L. Anceschi and J. Symons, 13–35. Basingstoke: Palgrave Macmillan.

Mazo, Jeffrey. 2010. *Climate Conflict: How Global Warming Threatens Security and What to Do About It.* Abingdon: Routledge.

Mazo, Jeffrey. 2011. The know-nothings. *Survival* 53 (6):238–248.

Meadows, Donella H., Dennis L. Meadows, Jørgen Randers, and William W. Behrens, III. 1972. *The Limits to Growth: A Report for the Club of Rome's Project on the Predicament of Mankind.* New York: Universe Books.

Meadows, Donella H., Dennis L. Meadows, and Jørgen Randers. 1992. *Beyond the Limits: Confronting Global Collapse, Envisioning a Sustainable Future.* Post Mills, VT: Chelsea Green.

Meadows, Donella H., Jørgen Randers, and Dennis L. Meadows. 2004. *Limits to Growth: The 30-Year Update.* White River Junction, VT: Chelsea Green.

Mearns, Robin, and Andrew Norton, eds. 2010. *Social Dimensions of Climate Change: Equity and Vulnerability in a Warming World.* Washington, DC: World Bank.

Merton, Robert K. 1973. *The Sociology of Science: Theoretical and Empirical Investigations.* Chicago: University of Chicago Press.

Mesa-Lago, Carmelo, ed. 1993. *Cuba after the Cold War.* Pittsburgh, PA: University of Pittsburgh Press.

Mesa-Lago, Carmelo, and Jorge F. Pérez-López. 2005. *Cuba's Aborted Reform: Socioeconomic Effects, International Comparisons, and Transition Policies.* Gainesville: University Press of Florida.

Mesarović, Mihajlo D., and Eduard Pestel. 1974. *Mankind at the Turning Point: The Second Report to the Club of Rome.* London: Hutchinson.

Mildner, Stormy-Annika, Gitta Lauster, and Wiebke Wodni. 2011. Scarcity and abundance revisited: a literature review on natural resources and conflict. *International Journal of Conflict and Violence* 5 (1):155–172.

Miller, Edward S. 2007. *Bankrupting the Enemy: The US Financial Siege of Japan before Pearl Harbor.* Annapolis, MD: Naval Institute Press.

Miller, Richard G. 2011. Future oil supply: the changing stance of the International Energy Agency. *Energy Policy* 39 (3):1569–1574.

Milly, P. C. D., Julio Betancourt, Malin Falkenmark, Robert L. Hirsch, Zbigniew W. Kundzewicz, Dennis P. Lettenmaier, and Ronald J. Stouffer. 2008. Stationarity is dead: Whither water management? *Science* 319:573–574.

Mitzen, Jennifer. 2006. Ontological security in world politics: State identity and the security dilemma. *European Journal of International Relations* 12 (3):341–370.

Monbiot, George. 2006. *Heat: How to Stop the Planet Burning.* London: Allen Lane.

Montgomery, David R. 2007. *Dirt: The Erosion of Civilizations.* Berkeley: University of California Press.

Moran, Daniel, and James A. Russell, eds. 2009. *Energy Security and Global Politics: The Militarization of Resource Management.* London, New York: Routledge.

Moriarty, Patrick, and Damon Honnery. 2009. What energy levels can the earth sustain? *Energy Policy* 37 (7):2469–2474.

Moriarty, Patrick, and Damon Honnery. 2011. *Rise and Fall of the Carbon Civilization: Resolving Global Environmental and Resource Problems.* London: Springer.

Moriarty, Patrick, and Damon Honnery. 2012. What is the global potential for renewable energy? *Renewable & Sustainable Energy Reviews* 16 (1):244–252.

Mouhot, Jean-François. 2011. Past connections and present similarities in slave ownership and fossil fuel usage. *Climatic Change* 105 (1):329–355.

Murphy, David J., and Charles A. S. Hall. 2011. Energy return on investment, peak oil, and the end of economic growth. *Annals of the New York Academy of Sciences* 1219:52–72.

Murphy, Pat. 2008. *Plan C: Community Survival Strategies for Peak Oil and climate Change.* Gabriola Island, BC: New Society Publishers.

Murray, James, and David King. 2012. Oil's tipping point has passed. *Nature* 481:433–435.

Natsios, Andrew. 2001. *The Great North Korean Famine.* Washington, DC: United States Institute of Peace Press.

Newell, Peter, and Matthew Paterson. 2010. *Climate Capitalism: Global Warming and the Transformation of the Global Economy.* Cambridge: Cambridge University Press.

Newman, Lenore, and Ann Dale. 2008. Limits to growth rates in an ethereal economy. *Futures* 40 (3):261–267.

Nordås, Ragnhild, and Nils Petter Gleditsch. 2009. IPCC and the Climate-Conflict Nexus. Paper presented to the 50th ISA Annual Convention, New York, February 15–18, 2009.

Nordhaus, Ted, and Michael Shellenberger. 2009. *Break Through: Why We Can't Leave Saving the Planet to Environmentalists.* Boston: Mariner.

Nordhaus, William D., and Joseph Boyer. 1999. *Warming the World: Economic Models of Global Warming.* Cambridge, MA: MIT Press.

Noreng, Øystein. 2006. *Crude Power: Politics and the Oil Market.* 2nd ed. London: I. B. Tauris.

Norgaard, Kari Marie. 2011. *Living in Denial: Climate Change, Emotions, and Everyday Life.* Cambridge, MA: MIT Press.

North, Douglass C. 1990. *Institutions, Institutional Change and Economic Performance.* Cambridge: Cambridge University Press.

North, Douglass C. 1991. Institutions. *Journal of Economic Perspectives* 5 (1):97–112.

Odell, Peter R. 2004. *Why Carbon Fuels Will Dominate the 21st Century's Global Energy Economy.* Brentwood, UK: Multi-Science.

Offer, Avner. 1989. *The First World War: An Agrarian Interpretation.* Oxford: Clarendon.

Offer, Avner. 2000. The blockade of Germany and the strategy of starvation, 1913–1918: An agency perspective. In *Great War, Total War: Combat and Mobilization on the Western Front, 1914–1918*, ed. R. Chickering and S. Förster, 169–188. Cambridge: Cambridge University Press.

Oki, Taikan, and Shinjiro Kanae. 2006. Global hydrological cycles and world water resources. *Science* 313:1068–1072.

Olson, Mancur. 1965. *The Logic of Collective Action: Public Goods and the Theory of Groups*. Cambridge, MA: Harvard University Press.

Ophuls, William. 1977. *Ecology and the Politics of Scarcity: Prologue to a Political Theory of the Steady State*. New York: Freeman.

Ophuls, William. 2011. *Plato's Revenge: Politics in the Age of Ecology*. Cambridge, MA: MIT Press.

Oreskes, Naomi, and Erik M. Conway. 2010. *Merchants of Doubt: How a Handful of Scientists Obscured the Truth on Issues from Tobacco Smoke to Global Warming*. New York: Bloomsbury.

Orlov, Dimitry. 2008. *Reinventing Collapse: The Soviet Example and American Prospects*. Gabriola Island, BC: New Society Publishers.

Orr, David W. 2009. *Down to the Wire: Confronting Climate Collapse*. Oxford: Oxford University Press.

Orr, David W., and David Ehrenfeld. 1995. None so blind: The problem of ecological denial. *Conservation Biology* 9 (5):985–987.

Ostrom, Elinor. 1990. *Governing the Commons: The Evolution of Institutions for Collective Action*. Cambridge: Cambridge University Press.

Ouellette, Judith A., and Wendy Wood. 1998. Habit and intention in everyday life: the multiple processes by which past behavior predicts future behavior. *Psychological Bulletin* 124 (1):54–74.

Owen, Nick A., Oliver R. Inderwildi, and David A. King. 2010. The status of conventional world oil reserves: Hype or cause for concern? *Energy Policy* 38 (8):4743–4749.

Padilla Dieste, Cristina. 2002. *Entre frijoles, papa y ají: La distribución de alimentos en Cuba*. Guadalajara: Universidad de Guadalajara.

Painter, James. 2011. *Poles Apart: The International Reporting of Climate Scepticism*. Oxford: Reuters Institute for the Study of Journalism.

Park, Philip Hookon. 2002. *Self-Reliance or Self-Destruction? Success and Failure of the Democratic People's Republic of Korea's Development Strategy of Self-Reliance "Juche"*. London and New York: Routledge.

Patterson, Wiliam P., Kristin A. Dietrich, Chris Holmden, and John T. Andrews. 2010. Two millennia of North Atlantic seasonality and implications for Norse colonies. *Proceedings of the National Academy of Sciences of the United States of America* 107 (12):5306–5310.

Pearson, Peter J. G., and Timothy J. Foxon 2012. A low carbon industrial revolution? Insights and challenges from past technological and economic transformations. *Energy Policy* 50: 117–127.

Peluso, Nancy Lee, and Michael Watts, eds. 2001. *Violent Environments*. Ithaca, NY: Cornell University Press.

Pérez-López, Jorge F. 1995. *Cuba's Second Economy: From Behind the Scenes to Center Stage*. New Brunswick: Transaction Publishers.

Perrings, Charles, and Bruce Hannon. 2001. An introduction to spatial discounting. *Journal of Regional Science* 41 (1):23–38.

Pertierra, Anna Cristina. 2011. *Cuba: The Struggle for Consumption*. Coconut Creek, FL: Caribbean Studies Press.

Peters, Glen P., Gregg Marland, Corinne Le Quéré, Thomas Boden, Joseph G. Canadell, and Michael R. Raupach. 2012. Rapid growth in CO2 emissions after the 2008 global financial crisis. *Nature Climate Change* 2 (1):2–4.

Pettit, Philip. 1997. *Republicanism: A Theory of Freedom and Government*. Oxford: Clarendon.

Pfeiffer, Dale Allen. 2006. *Eating Fossil Fuels: Oil, Food, and the Coming Crisis in Agriculture*. Gabriola Island, BC: New Society Publishers.

Pielke Jr., Roger A. 2007. *The Honest Broker: Making Sense of Science in Policy and Politics*. Cambridge: Cambridge University Press.

Piguet, Étienne, Antoine Pécoud, and Paul De Guchteneire, eds. 2011. *Migration and Climate Change*. Cambridge: Cambridge University Press.

Podobnik, Bruce. 2006. *Global Energy Shifts: Fostering Sustainability in a Turbulent Age*. Philadelphia: Temple University Press.

Polimeni, John M., Kozo Mayumi, Mario Giampietro, and Blake Alcott, eds. 2008. *The Jevons Paradox and the Myth of Resource Efficiency Improvements*. London: Earthscan.

Ponting, Clive. 2007. *A New Green History of the World: The Environment and the Collapse of Great Civilizations*. London: Vintage Books.

Pool, Robert. 1990. Struggling to do science for society. *Science* 248 (4956):672–673.

Poteete, Amy R., Marco A. Janssen, and Elinor Ostrom. 2010. *Working Together: Collective Action, the Commons, and Multiple Methods in Practice*. Princeton, NJ: Princeton University Press.

Putnam, Robert D. 2000. *Bowling Alone: The Collapse and Revival of American Community*. New York: Simon and Schuster.

Quiggin, John. 2008. Stern and his critics on discounting and climate change. *Climatic Change* 89 (3):195–205.

Quilley, Stephen. 2011. Entropy, the anthroposphere and the ecology of civilization: An essay on the problem of "liberalism in one village" in the long view. *Sociological Review* 59 (S1):65–90.

Quilley, Stephen. 2013. Degrowth is not a liberal agenda: Relocalization and the limits to low energy cosmopolitanism. *Environmental Values* 22 (2):261–286. Available electronically at http://www.whpress.co.uk/EV/papers/Quilley.pdf.

Raleigh, Clionadh, and Dominic Kniveton. 2012. Come rain or shine: An analysis of conflict and climate variability in East Africa. *Journal of Peace Research* 49 (1):51–64.

Raleigh, Clionadh, and Henrik Urdal. 2007. Climate change, environmental degradation and armed conflict. *Political Geography* 26 (6):674–694.

Ramsey, F. P. 1928. A mathematical theory of saving. *Economic Journal* 38 (152):543–559.

Randers, Jørgen. 2012. *2052: A Global Forecast for the Next Forty Years*. White River Junction, VT: Chelsea Green.

Ravetz, Jerome. 2004. The post-normal science of precaution. *Futures* 36 (3):347–357.

Reagan, Ronald. 1985. Second Inaugural Address, January 21, 1985.

Record, Jeffrey. 2009. *Japan's Decision for War in 1941: Some Enduring Lessons*. Carlisle, PA: Strategic Studies Institute.

Reuveny, Rafael. 2007. Climate change-induced migration and violent conflict. *Political Geography* 26 (6):656–673.

Rifkin, Jeremy. 2011. *The Third Industrial Revolution: How Lateral Power is Inspiring a Generation and Transforming the World*. New York: Palgrave Macmillan.

Roberts, Neil, Warren J. Eastwood, Catherine Kuzucuoğlu, Girolamo Fiorentino, and Valentina Caracuta. 2011. Climatic, vegetation and cultural change in the eastern Mediterranean during the mid-Holozene environmental transition. *Holocene* 21 (1):147–162.

Rockström, Johan, Will Steffen, Kevin Noone, Åsa Persson, F. Stuart Chapin, Eric F. Lambin, Timothy M. Lenton, et al. 2009. A safe operating space for humanity. *Nature* 461:472–475.

Rorty, Amélie Oksenberg. 1994. User-friendly self-deception. *Philosophy* 69: 211–228.

Rosendahl, Mona. 1997. *Inside the Revolution: Everyday Life in Socialist Cuba*. Ithaca, NY: Cornell University Press.

Rosset, Peter, and Medea Benjamin, eds. 1994. *The Greening of the Revolution: Cuba's Experiment with Organic Agriculture*. Melbourne: Ocean Press.

Rowe, Dorothy. 2010. *Why We Lie: The Source of our Disasters*. London: Fourth Estate.

Rozenberg, Julie, Stéphane Hallegatte, Adrien Vogt-Schilb, Olivier Sassi, Céline Guivarch, Henri Waisman, and Jean-Charles Hourcade. 2010. Climate policies as a hedge against the uncertainty on future oil supply. *Climatic Change* 101 (3):663–668.

Rubin, Jeff. 2009. *Why Your World Is About to Get a Whole Lot Smaller: Oil and the End of Globalization*. New York: Random House.

Rühl, Christof, Paul Appleby, Julian Fennema, Alexander Naumov, and Mark Schaffer. 2012. Economic development and the demand for energy: A historical perspective of the next 20 years. *Energy Policy* 50: 109–116.

Russell, Bertrand. 1912. *The Problems of Philosophy*. London: Williams and Norgate.

Russill, Chris. 2010. Stephen Schneider and the "double ethical bind" of climate change communication. *Bulletin of Science, Technology & Society* 30 (1): 60–69.

Sachs, Jeffrey D., and Andrew M. Warner. 1995. Natural Resource Abundance and Economic Growth. NBER Working Paper 5398.

Sachs, Jeffrey D., and Andrew M. Warner. 2001. Natural resources and economic development: the curse of natural resources. *European Economic Review* 45:827–838.

Sagan, Scott D. 1988. The origins of the Pacific War. *Journal of Interdisciplinary History* 28 (4):893–922.

Saloranta, Tuomo M. 2001. Post-normal science and the global climate change issue. *Climatic Change* 50 (4):395–404.

Scheffran, Jürgen, Michael Brzoska, Hans Günter Brauch, Peter Michael Link, and Janpeter Schilling, eds. 2012. *Climate Change, Human Security and Violent Conflict: Challenges for Societal Stability*. Heidelberg: Springer.

Schmitt, Carl. 1932. *Der Begriff des Politischen*. München, Leipzig: Duncker & Humblot.

Schneider, Stephen H. 1988. The greenhouse effect and the U.S. summer of 1988: Cause and effect of a media event? *Climatic Change* 13 (2):113–115.

Schneider, Stephen H. 2009. *Science as a Contact Sport: Inside the Battle to Save the Earth's Climate*. Washington, DC: National Geographic Press.

Schwartz, Peter, and Doug Randall. 2003. *An Abrupt Climate Change Scenario and its Implications for United States National Security: Imagining the Unthinkable*. Washington, DC: U.S. Department of Defense.

Schwekendiek, Daniel. 2011. *A Socioeconomic History of North Korea*. Jefferson, NC: McFarland & Company.

Shiva, Vandana. 2008. *Soil not Oil: Climate Change, Peak Oil, and Food Insecurity*. London: Zed.

Simmons, Matthew R. 2005. *Twilight in the Desert: The Coming Saudi Oil Shock and the World Economy*. Hoboken, NJ: John Wiley.

Smethurst, Richard J. 2007. *From Foot Soldier to Finance Minister: Takahashi Korekiyo, Japan's Keynes*. Cambridge, MA: Harvard University Press.

Smil, Vaclav. 2008a. *Energy in Nature and Society: General Energetics of Complex Systems*. Cambridge, MA: MIT Press.

Smil, Vaclav. 2008b. *Global Catastrophes and Trends: The Next Fifty Years*. Cambridge, MA: MIT Press.

Smil, Vaclav. 2010. *Energy Transitions: History, Requirements, Prospects*. Santa Barbara, CA: Praeger.

Smith, Kerri. 2011. We are seven billion. *Nature Climate Change* 1:331–335.

Smith, Laurence C. 2011. *The New North: The World in 2050*. London: Profile.

Sorrell, Steve. 2007. *The Rebound Effect: An Assessment of the Evidence for Economy-wide Energy Savings from Improved Energy Efficiency.* London: UK Energy Research Centre.

Sorrell, Steve, Richard Miller, Roger Bentley, and Jamie Speirs. 2010a. Oil futures: A comparison of global supply forecasts. *Energy Policy* 38 (9):4990–5003.

Sorrell, Steve, Jamie Speirs, Roger Bentley, Adam Brandt, and Richard Miller. 2010b. Global oil depletion: A review of the evidence. *Energy Policy* 38 (9):5290–5295.

Sorrell, Steve, Jamie Speirs, Roger Bentley, Richard Miller, and Erica Thompson. 2012. Shaping the global oil peak: A review of the evidence on field sizes, reserve growth, decline rates and depletion rates. *Energy* 37 (1):709–724.

Sovacool, Benjamin K. 2008. Valuing the greenhouse gas emissions from nuclear power: A critical survey. *Energy Policy* 36 (8):2950–2963.

Spratt, David, and Philip Sutton. 2008. *Climate Code Red: The Case for Emergency Action.* Melbourne: Scribe.

Steffen, Will, Åsa Persson, Lisa Deutsch, Jan Zalasiewicz, Mark Williams, Katherine Richardson, Carole Crumley, et al. 2011. The anthropocene: From global change to planetary stewardship. *Ambio* 40 (7):739–761.

Stern, Nicholas. 2007. *The Economics of Climate Change: The Stern Review.* Cambridge: Cambridge University Press.

Stevens, Paul. 2010. *The 'Shale Gas Revolution': Hype and Reality.* London: Chatham House.

Strotz, R. H. 1955. Myopia and inconsistency in dynamic utility maximization. *Review of Economic Studies* 23 (3):165–180.

Strumsky, Deborah, José Lobo, and Joseph A. Tainter. 2010. Complexity and the productivity of innovation. *Systems Research and Behavioral Science* 27 (5):496–509.

Swart, Rob, Lenny Bernstein, Minh Ha-Duong, and Arthur Petersen. 2009. Agreeing to disagree: uncertainty management in assessing climate change, impacts and responses by the IPCC. *Climatic Change* 92 (1–2):1–29.

Tainter, Joseph A. 1988. *The Collapse of Complex Societies.* Cambridge: Cambridge University Press.

Tainter, Joseph A. 2011. Resources and cultural complexity: implications for sustainability. *Critical Reviews in Plant Sciences* 30 (1):24–34.

Tainter, Joseph A., T. F. H. Allen, and T. W. Hoekstra. 2006. Energy transformations and post-normal science. *Energy* 31 (1):44–58.

Taylor, Graeme. 2008. *Evolution's Edge: The Coming Collapse and Transformation of our World.* Gabriola Island, BC: New Society Publishers.

Taylor, Henry Louis. 2009. *Inside el Barrio: A Bottom-Up View of Neighborhood Life in Castro's Cuba.* Sterling, VA: Kumarian Press.

Taylor, Shelley E., and Jonathon D. Brown. 1988. Illusion and well-being: A social psychological perspective on mental health. *Psychological Bulletin* 103 (2):193–210.

Theisen, Ole Magnus. 2008. Blood and soil? Resource scarcity and internal armed conflict revisited. *Journal of Peace Research* 45 (6):801–818.

Theisen, Ole Magnus, Helge Holtermann, and Halvard Buhaug. 2012. Climate wars? Assessing the claim that drought breeds conflict. *International Security* 36 (3):79–106.

Thompson, William R. 2006. Climate, water, and crises in the Southwest Asian Bronze Age. *Nature and Culture* 1 (1):88–132.

Tol, Richard S. J., and Sebastian Wagner. 2010. Climate change and violent conflict in Europe over the last millennium. *Climatic Change* 99:65–79.

Tonn, Bruce. 2007. The Intergovernmental Panel on Climate Change: A global scale transformative initiative. *Futures* 39 (5):614–618.

Trainer, Ted. 2007. *Renewable Energy Cannot Sustain a Consumer Society.* Dordrecht: Springer.

Trainer, Ted. 2012. A critique of Jacobson and Delucchi's proposals for a world renewable energy supply. *Energy Policy* 44: 476–481.

Turner, Graham, Steve Keen, and Franzi Poldy. 2011. Integrated assessment and scenarios. In *Resource Efficiency: Economics and Outlook for Asia and the Pacific*, 153–180. Bangkok: United Nations Environment Programme.

Turner, Graham M. 2008. A comparison of *The Limits to Growth* with 30 years of reality. *Global Environmental Change* 18 (3):397–411.

Twain, Mark. [1883] 2006. *Life on the Mississippi.* London: Folio.

UKERC. 2009. *Global Oil Depletion: An Assessment of the Evidence for a Near-Term Peak in Global Oil Production.* London: UK Energy Research Centre.

UN. 2011. *World Population Prospects: The 2010 Revision. Highlights and Advance Tables.* New York: United Nations.

Ur, Jason A. 2010. Cycles of civilization in Northern Mesopotamia, 4400–2000 BC. *Journal of Archaeological Research* 18 (4):387–431.

Urdal, Henrik. 2005. People vs. Malthus: Population pressure, environmental degradation, and armed conflict revisited. *Journal of Peace Research* 42 (4): 417–434.

Urry, John. 2011. *Climate Change and Society.* Cambridge: Polity.

Van de Graaf, Thijs. 2012. Obsolete or resurgent? The International Energy Agency in a changing global landscape. *Energy Policy* 48: 233–241.

Van de Graaf, Thijs, and Dries Lesage. 2009. The International Energy Agency after 35 years: Reform needs and institutional adaptability. *Review of International Organizations* 4 (3):293–317.

van den Bergh, Jeroen C. J. M. 2012. Effective climate-energy solutions, escape routes and peak oil. *Energy Policy* 46 (7):530–536.

Van den Hove, Sybille. 2007. A rationale for science-policy interfaces. *Futures* 39 (7):807–826.

Verbruggen, Aviel, and Mohamed Al Marchohi. 2010. Views on peak oil and its relation to climate change policy. *Energy Policy* 38 (10):5572–5581.

Victor, Peter A. 2008. *Managing without Growth: Slower by Design, Not Disaster.* Cheltenham, UK: Edward Elgar.

Vivoda, Vlado. 2009. Resource nationalism, bargaining and International Oil Companies: challenges and change in the new millennium. *New Political Economy* 14 (4):517–534.

Waisman, Henri, Julie Rozenberg, Olivier Sassi, and Jean-Charles Hourcade. 2012. Peak oil profiles through the lens of a general equilibrium assessment. *Energy Policy* 48: 744–753.

Walker, Brian, C. S. Holling, Stephen R. Carpenter, and Ann Kinzig. 2004. Resilience, adaptability and transformability in social-ecological systems. *Ecology and Society* 9 (2).

Ward, James D., Steve H. Mohr, Baden R. Myers, and Willem P. Nel. 2012. High estimates of supply constrained emissions scenarios for long-term climate risk assessment. *Energy Policy* 51: 598–604.

Warner, Ethan S., and Garvin A. Heath. 2012. Live cycle greenhouse gas emissions of nuclear electricity generation: Systematic review and harmonization. *Journal of Industrial Ecology* 16 (S1):73–92.

Warr, Benjamin, and Robert U. Ayres. 2010. Evidence of causality between the quantity and quality of energy consumption and economic growth. *Energy* 35 (4):1688–1693.

Warr, Benjamin, and Robert U. Ayres. 2012. Useful work and information as drivers of economic growth. *Ecological Economics* 73 (1):93–102.

Washington, Haydn, and John Cook. 2011. *Climate Change Denial: Heads in the Sand.* London and New York: Routledge.

WBGU. 2008. Climate Change as a Security Risk. London: Earthscan and German Advisory Council on Global Change (WBGU).

WBGU. 2011. *World in Transition: A Social Contract for Sustainability.* Berlin: German Advisory Council on Global Change (WBGU).

Weart, Spencer R. 2008. *The Discovery of Global Warming.* 2nd ed. Cambridge, MA: Harvard University Press.

Weart, Spencer R. 2012. The evolution of international cooperation in climate science. *Journal of International Organizations Studies* 3 (1):41–59.

Weisman, Avery D. 1972. *On Dying and Denying: A Psychiatric Study of Terminality.* New York: Behavioral Publications.

Weiss, Charles, and William B. Bonvillian. 2009. *Structuring an Energy Technology Revolution.* Cambridge, MA: MIT Press.

Weiss, Harvey. 2000. Beyond the Younger Dryas: Collapse as adaptation to abrupt climate change in ancient West Asia and the Eastern Mediterranean. In *Environmental Disaster and the Archaeology of Human Response,* ed. G. Bawden and M. Reycraft, 75–98. Albuquerque, NM: Maxwell Museum of Anthropology.

Weiss, Harvey, and Raymond S. Bradley. 2001. What drives societal collapse? *Science* 291:609–610.

Weiss, Harvey, M.-A. Courty, W. Wetterstrom, F. Guichard, L. Senior, R. Meadow, and A. Curnow. 1993. The genesis and collapse of Third Millennium North Mesopotamian Civilization. *Science* 261:995–1004.

Weitzman, Martin L. 2009. On modeling and interpreting the economics of catastrophic climate change. *Review of Economics and Statistics* 91 (1):1–19.

Weitzman, Martin L. 2011. Fat-tailed uncertainty in the economics of catastrophic climate change. *Journal of Environmental Economics and Policy* 5 (2):275–292.

Weitzman, Martin L. 2012. The Ramsey discounting formula for a hidden-state stochastic growth process. *Environmental Resource Economics* 53 (3): 309–321.

Welzer, Harald. 2011. *Climate Wars: Why People Will be Killed in the Twenty-First Century*. Cambridge: Polity.

Wen, Dale Jiajun. 2006. Peak oil preview: North Korea & Cuba. *Yes! Magazine*, Summer.

Westley, Frances, Per Olsson, Carl Folke, Thomas Homer-Dixon, Harrie Vredenburg, Derk Loorbach, John Thompson, et al. 2011. Tipping toward sustainability: Emerging pathways of transformation. *Ambio* 40 (7):762–780.

WFP/FAO/UNICEF. 2011. Special Report: WFP/FAO/UNICEF Rapid Food Security Assessment Mission to the Democratic People's Republic of Korea, March 24, 2011.

Whyte, Ian. 2008. *World without End? Environmental Disaster and the Collapse of Empires*. London: I. B. Tauris.

Wilkinson, Richard G. 1973. *Poverty and Progress: An Ecological Model of Economic Development*. London: Methuen.

Williams, James H., David Von Hippel, and Nautilus Team. 2002. Fuel and famine: Rural energy crisis in the DPRK. *Asian Perspective* 26 (1):111–140.

Willrich, Mason, and Melvin A. Conant. 1977. The International Energy Agency: An interpretation and assessment. *American Journal of International Law* 71 (2):199–223.

Woo-Cumings, Meredith. 2002. The Political Economy of Famine: The North Korean Catastrophe and Its Lessons. Tokyo: ADB Institute, Research Paper 31.

Woodbridge, Roy. 2004. *The Next World War: Tribes, Cities, Nations, and Ecological Decline*. Toronto: University of Toronto Press.

Worth, Roland H., Jr. 1995. *No Choice but War: The United States Embargo against Japan and the Eruption of War in the Pacific*. Jefferson, NJ: McFarland.

Wright, Gavin. 1986. *Old South, New South: Revolutions in the Southern Economy since the Civil War*. New York: Basic Books.

Wright, Gavin. 2006. *Slavery and American Economic Development*. Baton Rouge: Louisiana State University Press.

Wright, Julia. 2009. *Sustainable Agriculture and Food Security in an Era of Oil Scarcity: Lessons from Cuba*. London: Earthscan.

Wright, Ronald. 2004. *A Short History of Progress*. Edinburgh: Cannongate.

Yergin, Daniel. 1991. *The Prize: The Epic Quest for Oil, Money and Power*. New York: Free Press.

Yergin, Daniel. 2011. *The Quest: Energy, Security and the Remaking of the Modern World*. London: Allen Lane.

Zajko, Mike. 2010. Contested Science in the Global Warming Controversy. Graduate thesis, Faculty of Graduate Studies, University of Calgary, Calgary, Alberta.

Zajko, Mike. 2011. The shifting politics of climate science. *Society* 48 (6): 457–461.

Zerubavel, Eviatar. 2006. *The Elephant in the Room: Silence and Denial in Everyday Life*. Oxford: Oxford University Press.

Zhang, David D., Peter Brecke, Harry F. Lee, Yuan-Qing He, and Jane Zhang. 2007. Global climate change, war, and population decline in recent human history. *Proceedings of the National Academy of Sciences of the United States of America* 104 (49):19214–19219.

Zhang, David D., Harry F. Lee, Cong Wang, Baosheng Li, Qing Pei, Jane Zhang, and Yulun An. 2011. The causality analysis of climate change and large-scale human crisis. *Proceedings of the National Academy of Sciences of the United States of America* 148 (42):17296–17301.

Index